Chemical Toxicity Prediction
Category Formation and Read-Across

Issues in Toxicology

Series Editors:
Professor Diana Anderson, *University of Bradford, UK*
Dr Michael D Waters, *Integrated Laboratory Systems, Inc, N Carolina, USA*
Dr Martin F Wilks, *University of Basel, Switzerland*
Dr Timothy C Marrs, *Edentox Associates, Kent, UK*

Titles in the Series:
1: Hair in Toxicology: An Important Bio-Monitor
2: Male-mediated Developmental Toxicity
3: Cytochrome P450: Role in the Metabolism and Toxicity of Drugs and other Xenobiotics
4: Bile Acids: Toxicology and Bioactivity
5: The Comet Assay in Toxicology
6: Silver in Healthcare
7: *In Silico* Toxicology: Principles and Applications
8: Environmental Cardiology
9: Biomarkers and Human Biomonitoring, Volume 1: Ongoing Programs and Exposures
10: Biomarkers and Human Biomonitoring, Volume 2: Selected Biomarkers of Current Interest
11: Hormone-Disruptive Chemical Contaminants in Food
12: Mammalian Toxicology of Insecticides
13: The Cellular Response to the Genotoxic Insult: The Question of Threshold for Genotoxic Carcinogens
14: Toxicological Effects of Veterinary Medicinal Products in Humans: Volume 1
15: Toxicological Effects of Veterinary Medicinal Products in Humans: Volume 2
16: Aging and Vulnerability to Environmental Chemicals: Age-related Disorders and their Origins in Environmental Exposures
17: Chemical Toxicity Prediction: Category Formation and Read-Across

How to obtain future titles on publication:
A standing order plan is available for this series. A standing order will bring delivery of each new volume immediately on publication.

For further information please contact:
Book Sales Department, Royal Society of Chemistry, Thomas Graham House, Science Park, Milton Road, Cambridge, CB4 0WF, UK
Telephone: +44 (0)1223 420066, Fax: +44 (0)1223 420247
Email: booksales@rsc.org
Visit our website at www.rsc.org/books

Chemical Toxicity Prediction
Category Formation and Read-Across

Mark T. D. Cronin, Judith C. Madden, Steven J. Enoch, David W. Roberts
Liverpool John Moores University, Liverpool, UK
Email: m.t.cronin@ljmu.ac.uk

RSCPublishing

ISBN: 978-1-84973-384-7
ISSN: 1757-7179

A catalogue record for this book is available from the British Library

© M.T.D. Cronin, J.C. Madden, S.J. Enoch, D.W. Roberts 2013

All rights reserved

Apart from fair dealing for the purposes of research for non-commercial purposes or for private study, criticism or review, as permitted under the Copyright, Designs and Patents Act 1988 and the Copyright and Related Rights Regulations 2003, this publication may not be reproduced, stored or transmitted, in any form or by any means, without the prior permission in writing of The Royal Society of Chemistry or the copyright owner, or in the case of reproduction in accordance with the terms of licences issued by the Copyright Licensing Agency in the UK, or in accordance with the terms of the licences issued by the appropriate Reproduction Rights Organization outside the UK. Enquiries concerning reproduction outside the terms stated here should be sent to The Royal Society of Chemistry at the address printed on this page.

The RSC is not responsible for individual opinions expressed in this work.

Published by The Royal Society of Chemistry,
Thomas Graham House, Science Park, Milton Road,
Cambridge CB4 0WF, UK

Registered Charity Number 207890

For further information see our web site at www.rsc.org

From JCM
To JRM, CAM and JCM(Jr)

Preface

To address the societal problem of assessing the safety of hundreds of thousands of chemicals, without resorting to extensive animal testing, a paradigm shift was required in the way in which toxicity data were obtained. Computational tools to predict toxicity have been available for over 50 years, but there has previously been a reluctance to accept predictions from these models, particularly for regulatory purposes, with lack of model transparency often being identified as an issue. However, we are now in an era where toxicologists and computational modellers work together much more closely to resolve the problems posed in predictive toxicology, i.e. how to ensure the safety of chemicals for man and the environment. The mutual benefits of combining research efforts from diverse areas is now apparent and real progress is being made in both model development and acceptance.

This is a rapidly expanding field and the past few years have seen an increase in the numbers of publications in toxicology and the main toxicological conferences now routinely include modelling. There is also an increasing trend to see modelling at the heart of a toxicological study or project, e.g. to help identify chemicals to test, or to rationalise the results. Computational modelling has similarly benefitted from the input of toxicological expertise, for example in the development of consistent ontologies for new databases and the identification of key (modellable) steps within a toxicity pathway.

There are many well documented reasons for this shift in the way in which models are developed and used, not least the pressures of having to find solutions for European, and other, legislations e.g. REACH, the Cosmetics Regulation and others. Ethical pressures, financial and logistical constraints (not all the chemicals can be tested) as well as the adoption of 21st Century Toxicology, with the ultimate goal of moving away from the old animal-based paradigm of testing, have all played a role.

Issues in Toxicology No. 17
Chemical Toxicity Prediction: Category Formation and Read-Across
By Mark T. D. Cronin, Judith C. Madden, Steven J. Enoch, David W. Roberts
© M.T.D. Cronin, J.C. Madden, S.J. Enoch, D.W. Roberts 2013
Published by the Royal Society of Chemistry, www.rsc.org

This book addresses a specific area of computational toxicology that has seen remarkable growth in the past five years, namely the grouping of molecules – the so-called formation of categories – to allow for toxicity prediction from read-across. The general grouping approach has been shown to be transparent and easy to perform, making the process more accessible to toxicologists and more amenable for use in regulatory submissions. Whilst product development is not the primary aim of this technique, some of the information will be useful for designing safer new products or replacing existing toxic compounds with those that may be more benign. These new developments mean that for certain scenarios, i.e. well characterised chemicals and simpler endpoints, it could be argued that there is now little information that could be gained from testing that is not easily predicted from read-across; therefore obviating the need for *in vivo* testing for these compounds. For the more challenging endpoints, the situation is different and more work is required, not least identifying the key steps within the pathways, gathering and modelling data. Recently there has been a drive to develop a framework for organising the chemical and biological interactions that result in toxicity. This has led to the development of the Adverse Outcome Pathway approach, an important component of which is the identification of key steps that are amenable to modelling, e.g. by using read-across.

This book provides the background to the process of grouping for the purpose of read-across for toxicity prediction. It provides practical solutions for those wishing to perform read-across and identifies where more research and collaboration are needed and how this could be achieved. The concept of using this information within an Adverse Outcome Pathway is also described providing a framework for organising and using new information as it is generated. In Europe, over the time period in which this book was being written, (quite a long time as it turned out!) there have been several cycles of REACH submissions and the Cosmetics Regulation has been implemented. This has not only changed the way in which read-across for toxicity is perceived and utilised, but has also resulted in useful guidance, case studies and (often heated) debate on the subject. We have been fortunate that we have been guided by much high quality work from industry, academics, various parts of the European Commission, as well as the Organisation for Economic Co-operation and Development (OECD), not forgetting the Non Governmental Organisations. This book brings together the expertise developed recently in this area, providing greater insight and details on the process and application of read-across and its potential in developing Adverse Outcome Pathways. A history of the science, recent developments, practical, technical guidance and a philosophy for future developments are all presented.

Mark Cronin, Judith Madden and Steven Enoch
Liverpool John Moores University

Contents

Chapter 1   An Introduction to Chemical Grouping, Categories and
              Read-Across to Predict Toxicity   1
              M. T. D. Cronin

    1.1   Introduction – Ensuring the Safety of Exposure to Chemicals   1
        1.1.1 *In Silico* Predictions of Toxicity – Grouping, Category Formation and Read-Across   3
        1.1.2 *In Silico* Predictions of Toxicity – (Quantitative) Structure-Activity Relationships ((Q)SARs)   5
    1.2   Purpose of Category Formation and Read-Across   8
    1.3   History: From Structure-Activity to Grouping   9
    1.4   The Process of Category Formation and Read-Across   11
        1.4.1 Step 1 - Identification of the "Target" Chemical   11
            1.4.1.1 Identification of the Effect and/or Endpoint to Predict   13
        1.4.2 Step 2 - Identification of Similar Chemicals to the Target   13
        1.4.3 Step 3 - Obtaining Toxicity Data for the Grouping or Category   13
        1.4.4 Step 4 - Definition of the Category   14
        1.4.5 Step 5 - Prediction of Toxicity by Read-Across   14
        1.4.6 Step 6 - Documentation of the Prediction   14
            1.4.6.1 Accepting or Rejecting the Prediction   14
        1.4.7 Applying the Flow Chart Depicted in Figure 1.3   14
    1.5   Advantages and Disadvantages of Category Formation and Read-Across   15

Issues in Toxicology No. 17
Chemical Toxicity Prediction: Category Formation and Read-Across
By Mark T. D. Cronin, Judith C. Madden, Steven J. Enoch, David W. Roberts
© M.T.D. Cronin, J.C. Madden, S.J. Enoch, D.W. Roberts 2013
Published by the Royal Society of Chemistry, www.rsc.org

1.6	Uses of Read-Across and Category Formation - Current Literature	16
1.7	Key Literature and Guidance for the Regulatory Use of Read-Across	16
1.8	Aims of this Volume	17
1.9	Conclusions	20
Acknowledgement		20
References		20

Chapter 2   Approaches for Grouping Chemicals into Categories   30
S. J. Enoch and D. W. Roberts

2.1	Introduction	30
2.2	Methods of Defining Chemical Similarity Useful in Category Formation	31
2.3	Analogue Based Category	31
2.4	Common Mechanism of Action	32
	2.4.1 Structural Alerts for Developing Categories for Endpoints in Which Covalent Bond Formation is the Molecular Initiating Event	34
	2.4.2 Structural Alerts for Developing Categories for Endpoints in Which a Non-Covalent Interaction is the Molecular Initiating Event	35
2.5	Chemoinformatics	35
2.6	The Use of Experimental Data to Support the Development of Profilers for Chemical Category Formation	37
2.7	Adverse Outcome Pathways	38
2.8	Conclusions	39
Acknowledgements		39
References		40

Chapter 3   Informing Chemical Categories through the Development of Adverse Outcome Pathways   44
K. R. Przybylak and T. W. Schultz

3.1	Introduction	44
3.2	The Structure of the AOP	46
	3.2.1 Development of the AOP	46
	3.2.1.1 Identification of the Adverse Effect	48
	3.2.1.2 Definition of the Molecular Initiating Event (MIE)	48
	3.2.1.3 Recognition of Key Events Leading to the Adverse Effect	49

Contents   xi

		3.2.2	The Assessment of the AOP	49
	3.3		Harmonised Reporting and Recording of an AOP	50
	3.4		Use and Benefits of an AOP	51
		3.4.1	Developing Chemical Categories Supported by an AOP	51
		3.4.2	General Applications of AOP for Regulatory Purposes	53
	3.5		A Case Study: Developing a Chemical Category for Short-Chained Carboxylic Acids Linked to Developmental Toxicity	53
		3.5.1	Overview of Developmental Toxicity	53
		3.5.2	Valproic and Other Short-Chained Carboxylic Acids as Developmental Toxicants	55
		3.5.3	AOP for Short-Chained Carboxylic Acids as Developmental Toxicants to Organisms in Aquatic Environments	57
		3.5.4	A Case Study Using Carboxylic Acid Chemical Categories to Evaluate Developmental Hazard to Species in Aquatic Environments	58
	3.6	Conclusions		67
	Acknowledgements			67
	References			67

Chapter 4 Tools for Grouping Chemicals and Forming Categories   **72**
J. C. Madden

4.1	Introduction	72
4.2	Reasons for Grouping Compounds	73
4.3	The OECD QSAR Toolbox	74
	4.3.1 The Workflow of the Toolbox	75
4.4	The Hazard Evaluation Support System (HESS)	87
4.5	Toxmatch	88
4.6	Toxtree	91
4.7	AMBIT	92
4.8	Leadscope	92
4.9	Vitic Nexus	93
4.10	ChemSpider	93
4.11	ChemIDPlus (Advanced)	94
4.12	Analog Identification Methodology (AIM)	94
4.13	Use of Computational Workflows in Read-Across	95
4.14	Conclusions	95
Acknowledgements		97
References		97

| Chapter 5 | Sources of Chemical Information, Toxicity Data and Assessment of Their Quality | 98 |

J. C. Madden

		5.1	Introduction	98
		5.2	Data Useful for Category Formation and Read-Across	99
		5.3	Sources of Data	101
			5.3.1 In-house Data Sources	101
			5.3.2 Public Data Sources	102
		5.4	Strategies for Data Collection	104
		5.5	Data Quality Assessment	108
			5.5.1 Accurate Identification and Representation of Chemical Structure	108
			5.5.2 Quality Assessment of Computationally-Derived Chemical Descriptors	113
			5.5.3 Quality Assessment of Experimentally Derived Data	113
			5.5.4 Guidance and Tools for Data Quality Assessment	117
			5.5.5 Alternative Assessment Schemes	121
			5.5.6 Problems with Assessment	122
		5.6	Conclusions	123
		Acknowledgements		124
		References		124

| Chapter 6 | Category Formation Case Studies | 127 |

S. J. Enoch, K. R. Przybylak and M. T. D. Cronin

		6.1	Introduction	127
		6.2	Mechanism-based Case Studies	128
			6.2.1 Case Study One: Category Formation for Ames Mutagenicity	128
			6.2.2 Case Study Two: Category Formation for Skin Sensitisation	132
			6.2.3 Case Study Three: Category Formation for Aquatic Toxicity	135
			6.2.4 Case Study Four: Category Formation for Oestrogen Receptor Binding	137
			6.2.5 Case Study Five: Category Formation for Repeated Dose Toxicity	144
		6.3	Similarity-based Case Studies	149
			6.3.1 Case Study Six: Category Formation for Teratogenicity	150
		6.4	Conclusions	152
		Acknowledgements		152
		References		152

Chapter 7   Evaluation of Categories and Read-Across for Toxicity Prediction Allowing for Regulatory Acceptance   155
M. T. D. Cronin

7.1	Introduction	155
7.2	Assigning Confidence to the Robustness of a Category	156
7.3	Assigning Confidence to the Read-Across Prediction	157
	7.3.1 Weight of Evidence to Support a Prediction	159
7.4	Reporting of Predictions	159
	7.4.1 Tools for Category Description and Prediction	160
7.5	Regulatory Use of Predictions	160
7.6	Training and Education	165
7.7	Conclusions	165
	Acknowledgement	166
	References	166

Chapter 8   The State of the Art and Future Directions of Category Formation and Read-Across for Toxicity Prediction   168
M. T. D. Cronin

8.1	Introduction		168
8.2	Current State of the Art		169
	8.2.1	High Quality Tools Available to Assist the User	169
	8.2.2	Understanding of the Processes of Grouping and Read-Across	169
	8.2.3	Growth in Toxicological Databases to Support Read-Across	170
	8.2.4	Status of Profilers for Category Formation	170
	8.2.5	Prediction and Profiling of Metabolism	172
	8.2.6	Training and Education	173
8.3	The Future for Category Formation and Read-Across		173
	8.3.1	Upkeep and Maintenance of Current Tools	173
	8.3.2	Development of New Profilers	174
	8.3.3	Incorporation of Toxicokinetic Information	174
	8.3.4	Using the Adverse Outcome Pathway (AOP) Concept to Support Category Formation	175
	8.3.5	Global Co-ordination of the Development of Adverse Outcome Pathways	175
	8.3.6	Better Use of New Toxicological Data and Information Sources	175
	8.3.7	Data Quality Assessment	176
	8.3.8	Confidence in Predictions	176

	8.3.9 Acceptance of Predictions for Regulatory Purposes	176
	8.3.10 Education and Training	177
8.4	Conclusions	177
Acknowledgement		177
References		177

Subject index   **180**

CHAPTER 1

An Introduction to Chemical Grouping, Categories and Read-Across to Predict Toxicity

M. T. D. CRONIN

School of Pharmacy and Chemistry, Liverpool John Moores University, Byrom Street, Liverpool L3 3AF, England
E-mail: m.t.cronin@ljmu.ac.uk

1.1 Introduction – Ensuring the Safety of Exposure to Chemicals

Modern society requires safe chemicals. However, nothing is without risk and there is increasing pressure to identify hazardous chemicals and replace them with those that are more benign. In order to ensure the well-being of their population and the environment, governments enforce legislation to determine the effects of chemicals and ensure that every day accidental or occupational exposure will not cause harm. This is desirable for all substances that man comes into contact with, or that may be released into the environment, whether the substance is in foods, medicines, pesticides, fertilisers, or cosmetic ingredients (amongst many other types of chemicals that are in use). Different regulations are applicable to each type of chemical associated with a particular use.

In order to determine the risk associated with the use of a chemical, a certain amount of information is required. Firstly, a means of defining risk is a pre-requisite. In this context, risk is a function of the intrinsic hazard of a chemical and the exposure. Considering hazard, this can be considered as the ability to cause harm to a species, be that organisms that are deliberately exposed to the chemical or a non-target (for instance, environmental) species. Exposure can be simplistically considered as the quantity of a chemical to which the target and non-target species are exposed.[1]

Within the current definition of risk, information is required regarding the hazards of chemicals; this is provided by the science of toxicology. Assessing the toxicity of chemicals involves determining what the harmful effects of a chemical may be, *i.e.* toxicity to particular organs, effects to the skin, lethality, tumour promotion and countless others. Assessment normally involves testing for these effects and being able to use the test results in a manner that is protective of man and the environment. The tests and the information they provide need to be scientifically credible and satisfy the needs of the manufacturer, government or regulatory agency that has to interpret them and, ultimately, the user or consumer for whom safety must be assured. The information must be reliable, trustworthy and protective, *i.e.* precautionary. The need to determine the hazard of a chemical has resulted in the toxicological testing of chemicals for a wide range of specific effects, *e.g.* the ability to promote tumours. These effects can then be reported and interpreted to identify hazard. The most accepted paradigm for the identification of the majority of toxic effects has been the use of animal testing, through a series of standardised assays. However, the use of animals to identify hazard has received much criticism as being unethical, difficult to extrapolate results and findings to humans, costly and not always capable of identifying subtle or idiosyncratic toxicities.[2] Therefore, for decades, alternatives to animal testing have been sought. Amongst these are the so-called computational, or *in silico*, models which attempt to draw conclusions regarding the toxicity of a chemical from existing knowledge and/or its chemical structure. It is a selection of these techniques, those involving grouping similar chemicals together and reading across (or interpolating) activity, that form the focus of this volume.

With regard to exposure, a number of issues must be considered. The first is how much of the material is the organism in question exposed to? Also of importance is the time period over which the organism will be exposed, the route and manner of administration, *i.e.* the formulation (that may affect uptake). Consideration must also be given to whether local, *i.e.* at the site of exposure (if applicable), or systemic effects are of concern. Assessment of exposure therefore requires appreciation of uptake and bioavailability within the organism. A key principle to remember is that if there is no exposure to a chemical, or it is at a level below that which can cause harm (as defined by the toxicological assessment), there will be no risk.

In silico or computational toxicity prediction methods cover a very wide range of techniques and approaches, some of which are described in Sections 1.1.1 and 1.1.2. However, the main focus of this volume is to describe in detail category formation and read-across.

1.1.1 *In Silico* Predictions of Toxicity – Grouping, Category Formation and Read-Across

Similar objects tend to have similar properties. Applied to chemistry, this means that for chemicals that can be classed as being similar to other chemicals, we can understand and predict their properties without the need for testing. This fundamental concept has been applied to the prediction of properties and harmful effects of compounds for decades. Thus, being able to form groups of similar compounds (also called categories) becomes a powerful approach. If a compound belongs to a group of compounds with a well categorised toxicological profile, it can be possible to interpolate its activity. These interpolations, (predictions) of toxicity may, when utilised properly, provide hazard information that can be used in the assessment procedure described above. The process of prediction is termed "read-across" as it assumes that activities, toxicities or properties can be read across between compounds within a category.

Two hypothetical examples of read-across are provided in Figures 1.1 and 1.2 — these use data obtained from the OECD QSAR Toolbox version 3.1 (see Section 4.3 for more details). In the first example, Figure 1.1, a read-across prediction of *Salmonella typhimurium* gene mutation is made for 2-(3-ethylphenyl)oxirane. No mutagenicity data are available for this chemical. However, *S. typhimurium* gene mutation data are available for four closely related chemicals — termed analogues 1–4. These chemicals are considered "similar" as they all contain an aromatic ring, an epoxy group and limited alkyl substitution. The epoxy group allows the chemical to act as a direct acting electrophile by the S_N2 mechanism.[3] All the analogues to the target chemical are positive in the *S. typhimurium* gene mutation assay, they share the same structural features to the target, hence the read-across prediction for the target is that it will also share the same mechanism and be positive in this assay. This is therefore an example of a "qualitative" read-across.

The second hypothetical example is shown in Figure 1.2. This is a quantitative read-across in that a prediction is made for the acute fish toxicity of 3,4-dimethyl-1-pentanol. 96 hour LC_{50} values to the fathead minnow (*Pimephales promelas*) were retrieved for six analogues. These analogues are similar in that they are all simple saturated aliphatic molecules with a hydroxy group. As such they are assumed to act by the same mechanism of action — termed non-polar narcosis — and a good relationship is expected with the logarithm of the octanol-water partition coefficient (log P).[4] Figure 1.2 actually demonstrates the development of a local Quantitative Structure-Activity Relationship (QSAR), the line of best fit between toxicity and log P has the following equation:

$$\text{Toxicity} = 0.963 \log P + 0.769 \tag{1}$$

Where:
Toxicity is the inverse logarithm of the 96 hour LC_{50} values to *Pimephales promelas* (millimoles per litre).

Target Chemical	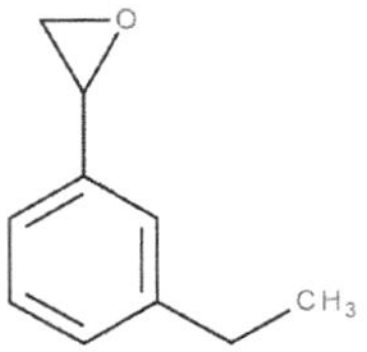 *[structure: 3-ethylphenyl oxirane with CH₃]*	Read-across prediction is *Salmonella typhimurium* gene mutagen
Analogue 1	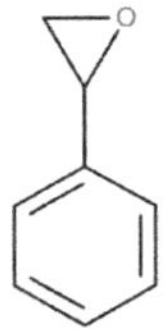 *[structure: phenyl oxirane]*	*Salmonella typhimurium* gene mutagen
Analogue 2	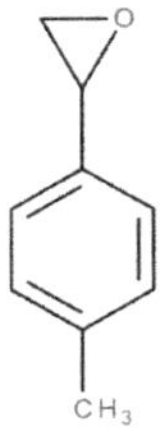 *[structure: 4-methylphenyl oxirane]*	*Salmonella typhimurium* gene mutagen
Analogue 3	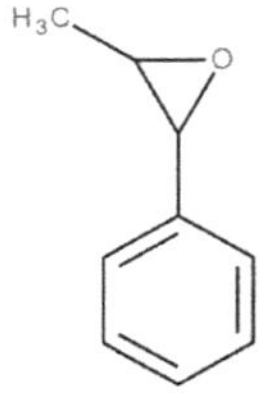 *[structure: 2-methyl-2-phenyl oxirane]*	*Salmonella typhimurium* gene mutagen
Analogue 4	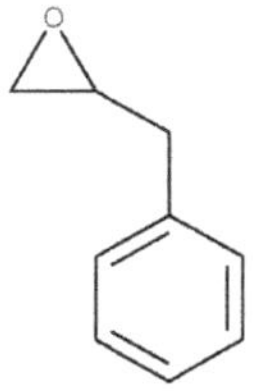 *[structure: benzyl oxirane]*	*Salmonella typhimurium* gene mutagen

Figure 1.1 An example of a hypothetical read-across for the mutagenicity of 2-(3-ethylphenyl)oxirane from four analogues with experimental activity retrieved from the OECD QSAR Toolbox.

Equation (1) has very good statistical fit (the correlation coefficient is 0.99). The target chemical has a log P value of 2.17, hence toxicity is calculated to be 2.86 (log units).

The purpose of these grouping and read-across techniques is described in detail below (Section 1.2), the goal being to predict the effects of compounds directly from chemical structure. The area of "predictive" toxicology, including computational techniques, has seen rapid growth and development

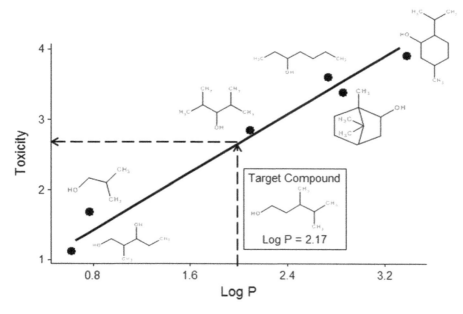

Figure 1.2 An example of a hypothetical read-across for the fish acute toxicity of 3,4-dimethyl-1-pentanol from six analogues with experimental toxicity values retrieved from the OECD QSAR Toolbox.

over several decades, fuelled by the desire to know more about the properties of chemicals. The history of this area is described in Section 1.3. Read-across is one of the most simplistic approaches to predict toxicity. There are several other levels of computational techniques, increasing in complexity and (probably) sophistication, all of which are well established, which can be applied to predict toxicity and other endpoints; these are summarised in Section 1.1.2.

It is true to say that read-across has grown in popularity due to the realisation that other types of modelling are not likely to be predictive for some sub-acute endpoints, especially those associated with repeat dose toxicity. Some of the key phrases and concepts with regard to read-across are defined in Table 1.1.[5,6] Whilst read-across is simplistic and, in theory at least, easy to apply, there are a number of drawbacks; such advantages and disadvantages of these approaches are described in Section 1.5.

1.1.2 *In Silico* Predictions of Toxicity – (Quantitative) Structure-Activity Relationships ((Q)SARs)

Read-across is a fundamental and empirical approach to predict activity. There is also a wide variety of techniques where models have been developed from larger groups of data, which use more detailed descriptors of molecules. These techniques are broadly termed Quantitative Structure-Activity

Table 1.1 Basic definitions and concepts for read-across. Please note that more formal definitions may be available from the OECD[5] and ECHA.[6]

Term	Definition
In Silico (or Computational) Toxicology	Computer-based methods to predict the toxicity or other properties of a substance directly from its structure.
Computational Chemistry	Application of computational techniques to chemistry to perform calculations on chemical structure. Methods can range from relatively simple to high level *ab initio* calculations.
Chemoinformatics	Linkage between chemistry and informatics to store chemical structures and related data, create knowledge, mine data and develop models for activities and properties.
Structural Alerts	Fragments of a molecule (usually identified by human experts or, occasionally, artificial intelligence) associated with a particular activity. In the context of this volume this implies fragments associated with toxicity that are normally supported by mechanistic interpretation.
Structure-Activity Relationships (SARs)	The formation of qualitative relationships between aspects of chemical structure (typically a functional group or 2-D/3-D arrangement of functional groups) and activity.
Chemical Grouping or Category Formation	The grouping of similar chemicals together to form a category.
Read-Across	Interpolation of activities or properties within a group or category of similar chemicals to make a prediction of an activity from known data.
Similarity (of Chemicals)	The degree of similarity between two or more objects. In this volume it relates specifically to the similarity between chemical structures. There are a variety of means of assessing chemical similarity including comparing molecules that contain the same functional group(s) or physico-chemical properties or using calculated measures of similarity.
Profiler	In the context of this volume a profiler is a collection of structural alerts that can be used to profile a chemical in order to assist in grouping.
Analogues	Analogues are structurally similar chemicals, normally having the same functional group(s) and differing only in carbon chain length (or another very simple molecular property).
Regulatory Acceptance	The process by which a read-across prediction can be documented and presented to provide sufficient evidence to allow for a regulatory decision to be made about that chemical.

Table 1.1 (*Continued*)

Term	Definition
Adverse Outcome Pathway (AOP)	A broad framework to organise information from the exposure of a chemical to adverse effects to an organism, population or ecosystem. It is defined by the molecular initiating events, a series of key events and the adverse outcome. It includes mechanistic information and allows chemistry (in terms of the initiating event) to be linked directly to an adverse effect.
Molecular Initiating Event (MIE)	The starting point of a toxicity pathway and incorporated into the Adverse Outcome Pathway concept. The MIE describes the interaction between the chemical and the biological system(s) that perturbs the biochemical pathway.

Relationships (QSARs) and include aspects of statistical modelling. Good overviews of the types of QSAR models widely used are available.[7,8]

Whilst this book focuses on category formation it is wrong to exclude it from the other techniques of toxicity and property prediction, *i.e.* SARs, QSARs and expert systems. Read-across can be considered to be a simplistic form of QSAR analysis. Indeed, within a category simple QSARs may be formed, should sufficient data be available. Some of the basic definitions relevant to these quantitative approaches are provided in Table 1.2.

Table 1.2 Basic Definitions and Concepts for QSAR and Expert Systems.

Term	Definition
Quantitative Structure-Activity Relationship (QSAR)	A mathematical model that relates (usually statistically) the activity or potency of a series of chemicals to physico-chemical properties or descriptors of the chemicals.
Expert System	A general term relating to software that may automate SAR or QSAR approaches, allowing for ease of use to make predictions.
Molecular Modelling	The computational modelling of molecules and their interactions *e.g.* receptor–ligand binding to provide graphical displays and provide the basis for subsequent calculations.
Molecular Descriptors	Measured or calculated structural or physico-chemical properties of a molecule that may be related to activity in a QSAR.

1.2 Purpose of Category Formation and Read-Across

The grouping of similar objects together, forming patterns and attempting to make a rational and reasoned interpretation is a sign of an intelligent and numerate mind. The human brain has instinctive capabilities to understand the relationship between objects and develop knowledge. For many years attempts have been made to capture both the processes of knowledge gathering and the knowledge. It is easier to recreate the knowledge than the process of achieving it.

With regard to chemical structure, it is almost instinctive to begin to group "similar" molecules together. Medicinal chemists have for many years been familiar with the concept of identifying similar molecules in terms of pharmacological activity. More recently, development of this concept has been necessitated by the global need to assess the properties and safety of new and, more urgently, existing chemicals. Most significantly, these methods will be used to identify similar chemicals with regard to toxicity. The purpose therefore is to provide information for chemicals where it is missing. These missing data, or data gaps, are most prevalent for existing industrial chemicals, particularly those produced in low tonnage.

The purpose of forming categories and performing read-across extends to the prediction of the effects of chemicals to humans and the environment. It should not be overlooked that grouping and read-across may also be used to predict other effect data, *e.g.* physico-chemical properties (log P) and many others and properties relating to ADME (such as skin permeability).

The use of category formation and read-across as a technique is seeing growth for a number of reasons, including the following:

- A realisation that many chemicals have missing toxicological and physico-chemical data (data gaps) and that these may be crucial for understanding the risk posed by chemicals.
- Chemicals legislation that has forced the need for rapid non-test methods to assess chemical safety.
- Acceptance (or at least partial acceptance) in the past by regulatory agencies of read-across to provide information for regulatory submissions.
- The development of new software *e.g.* the OECD QSAR Toolbox as well as easier access to toxicological databases and methods to assess and determine the similarity of chemicals.
- Category formation and read-across is seen as a clear, simple and transparent technique to make *in silico* predictions, thus increasing use by allowing for ease of regulatory acceptance.
- Traditional (Q)SAR methods have often performed poorly for complex toxicological endpoints *e.g.* repeat dose toxicity and reproductive effects in humans, as well as chronic toxicity in environmental species. Even fundamental properties such as water solubility have proved difficult to

predict. Read-across has been shown to provide a more reasonable solution to these problems than (Q)SAR.
- Read-across can allow for predictions to be made with a small number of data (even when there may only be one data point, should that be of high quality and the category very robust).

1.3 History: From Structure-Activity to Grouping

It is almost impossible to provide a detailed history of the use of category formation for toxicity prediction. The reason for this is that little of it is documented in the public literature. However, it is certain that *ad hoc* grouping and read-across has been undertaken for decades in areas such as medicinal chemistry, toxicity and ADME property prediction. Often this was simply termed structure-activity, or the development of structure-activity relationships.

Considering the literature available, what is certain is that there are only a handful of publications (possibly fewer than twenty) dealing with read-across for toxicity prediction prior to 2005 — see for instance the publications noted in Table 1.3.[3,9-95] It is also true that since then the number of publications dealing with category formation and read-across is growing rapidly year-on-year with no sign of a cessation of growth. There is no coincidence in these dates or growth in interest.

Structure-activity has been a cornerstone of understanding and rationalising the toxic effects of chemicals for the best part of a century. For instance, by the 1950s there was a clear appreciation of the relationship between carcinogenic activity of molecules and their shape, structure and properties (cf. Lacassagne *et al.*)[96] with much fundamental work being performed several decades before then. These early ventures into SAR were not captured electronically until computational technology caught up with the science in the (late) 1980s and early 1990s. Examples of computational methods for predicting toxicity include: the DEREK system[97] which later became DEREK for Windows and more recently DEREK NEXUS, and the United States Environmental Protection Agency's (US EPA's) ECOSAR[98] and Oncologic systems.[99] Other good examples of the application of chemistry to explain toxicity were the books by Dupuis and Benezra[100] and Lien.[101] The book by Dupuis and Benezra[100] was remarkably visionary and is often overlooked, it was probably a decade before its contents were taken up again. Basically, the book sets out the SAR behind skin sensitisation (in terms of protein reactivity) which is still the basis of grouping in this area to this day. The other well known and still used example of toxicological SAR from this era was the mutagenic "supermolecule" devised by Ashby and Tennant.[102]

The 1990s saw the growth in computational technology that allowed SAR knowledge to become common in the workplace and eventually on the desktop. At this time however, grouping and read-across (particularly for regulatory purposes) were undertaken, but little recognised, at least in the European Union. Interest in computational methods to predict toxicity grew following the adoption in 2001, by the European Commission, of a White Paper setting out the strategy for a future Community Policy for Chemicals.[103] The main objective of the new chemical strategy was to ensure a high level of protection for human health and the environment. One of the consequences was to require more information regarding the safety and potential harmful effects of chemicals. In the absence of extant toxicological data, there appeared to be few options to obtain such data other than testing or the use of predictions from computational toxicology. Interestingly, as the legislation progressed into regulation, use of all non-test data through the application of Integrated Testing Strategies (ITS), as well as exposure-based waiving, also came to the fore. With regard to computational testing, a group of approximately 60 industry, regulatory and academic scientists met in Setubal, near Lisbon, Portugal in March 2002 to set out the approaches for what became the framework for the regulatory use and acceptance of (Q)SARs.[104] At the time of the Setubal Workshop, reviews of the regulatory use of QSAR made little mention of the existing category formation and read-across methods.[105,106] Therefore, it can be concluded that the subsequent ten years were the realisation of this approach and its routine application.

Key to the acceptance of predictions from (Q)SARs and read-across was the definition of the "Setubal" Principles for the Validation of (Q)SARs (from the Workshop), which later developed and transformed into the OECD Principles for the Validation of (Q)SARs — which have now gained very broad acceptance.[107] The OECD Validation Principles are the subject of much debate and considerable guidance concerning their application is available from the Organisation for Economic Cooperation and Development (OECD), the European Chemicals Agency (ECHA) as well as industry and other bodies. They allow a user to document a prediction as evidence to support it, and provide the regulatory agencies a framework to enable acceptance of a prediction.

Another recommendation from the Setubal Workshop was the development of what was originally termed a "Decision Support System". The concept for this was a system to allow for databases and models to be brought together to support risk assessment decisions. A work programme was put in place to develop this tool, the emphasis being to develop software that would allow for the grouping of chemicals and application of read-across. As such, the first version of the OECD QSAR Application Toolbox was released in March 2008, with later versions of the (renamed) OECD QSAR Toolbox subsequently being released. The OECD QSAR Toolbox is described elsewhere (Section 4.3) but was one of the key factors in the uptake and use of category formation and read-across. There are many reasons for this, but they include the timing of its release, the quality of the databases and profilers, its free availability and the fact is is extremely well supported by documentation and training materials.

The incredible "success" of read-across, in terms of its uptake and use, was illustrated by reports issued by ECHA relating to data submitted in 24,560 registration dossiers between 1st June 2008 and 28th February 2011. In particular, a report from ECHA[108] details the use of alternatives to animal testing in these dossiers. The surprising fact was that between 20–30% of dossiers contained "read-across" estimates for all the toxicity endpoints considered. The ECHA report and the information contained within were analysed by Spielmann *et al.*[109] and whilst the data require more analysis and interpretation, it is clear that read-across had been widely attempted.

The recent uptake of read-across follows a number of other trends, particularly the response to the European Union's Cosmetic Regulation banning animal testing on new cosmetics ingredients.[110] Scientifically, new ways of providing information on toxicity are required and are being developed. For instance in the USA, an immense effort is being undertaken under the auspices of the "Toxicology in the 21st Century" (Tox21) initiative (http://epa.gov/ncct/Tox21/). This effort, combined with the use of Adverse Outcome Pathways (AOPs — see Chapter 3) to organise information, may provide the framework to identify and justify chemical categories. This fundamental grounding of chemistry approaches to mechanisms and modes of action of toxicity is one of the cornerstones of regulatory acceptance and provides a means to develop robust categories.

1.4 The Process of Category Formation and Read-Across

There are a number of distinct and definable steps to grouping chemicals and making a read-across prediction. The general process of category formation and read-across is summarised in Figure 1.3. More detail on this process is given below and in the relevant chapters. At the outset some factors must be appreciated, namely that implementation of the process requires expertise and must be considered as a knowledge-based and subjective approach; the prediction must be performed on a case-by-case basis taking account of the chemical itself, its use and importantly the endpoint being modelled. The same compound may quite feasibly be in different categories for different endpoints.

The following sections refer to Figure 1.3. The flow chart presented in Figure 1.3 is inspired by and relies heavily on that published by the OECD[5] and ECHA[6], as well as work by other authors. In particular it replicates the work flow represented by the OECD QSAR Toolbox – it should not be considered as being entirely the current author's own creation!

The following sections should be read in conjunction with Figure 1.3. There are six "steps" to the flow chart represented by the solid numbered boxes at the centre of the flow chart — these relate specifically to the work flow around which the OECD QSAR Toolbox is developed (Figure 4.1).

1.4.1 Step 1 - Identification of the "Target" Chemical

The chemical substance for which the prediction is required is termed the "target" chemical. It is assumed for the purposes of this introduction that this

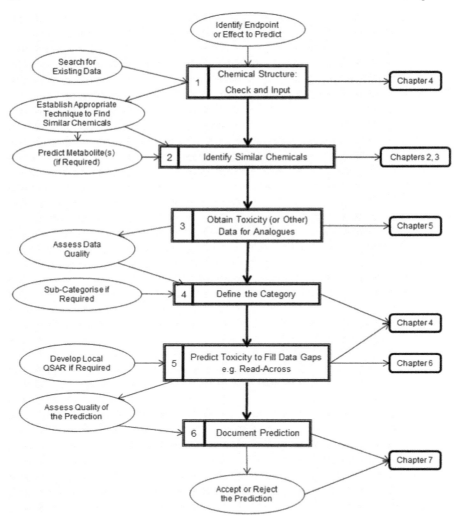

Figure 1.3 Flow chart of the grouping and read-across process to predict toxicity.

is a single chemical substance. Predictions for mixtures are more complex, but still achievable if the individual components are considered. At this point, it is essential that the target structure is defined definitively, including recognising and documenting stereochemistry and tautomers. It is normal for the structure to be considered as a neutral molecule, with no salt. However, the possible effects of ionisation may need to be borne in mind with regard to the bioavailability of a substance.

1.4.1.1 *Identification of the Effect and/or Endpoint to Predict*

The endpoint to be predicted for the target chemical needs to be identified. This may appear to be a relatively trivial task, *e.g.* mutagenicity, but in reality is dependent on the available data. Therefore, for mutagenicity, there may be different relevant data from different assays *e.g.* Ames, Chromosomal mutation *etc.* At this point, it is worth searching for existing data for the target chemical. There are many possible databases of toxicological and other information (*e.g.* eChemPortal). The existing data for the target chemical should be evaluated and assessed for their utility, (Chapter 5 covers aspects of finding and assessing data). Whilst it is obvious to state, should the existing data be suitable for purpose, there is no requirement to undertake further predictive or experimental studies.

1.4.2 Step 2 - Identification of Similar Chemicals to the Target

To form a group or category of similar chemicals, suitable criteria for assessing similarity are required. Such criteria are discussed in Chapter 2. They can range from a "chemist's" viewpoint *e.g.* congeneric series, a "chemoinformatician's" viewpoint *e.g.* molecular similarity or a "toxicologist's" viewpoint *e.g.* chemicals with a similar mechanism or mode of action. Thus for the same chemical, different groups or categories can be formed. The "best" or "most appropriate" group will depend on the nature of the chemical, available data for read-across and the endpoint to be modelled — please refer to Chapters 2 and 3.

Once an approach to assessing similarity has been established, chemicals with similar structures can be identified. There are different approaches that may be applied at this point. The most "liberal", and probably most favoured, is to search inventories of chemical structures regardless of whether they are associated with toxicity data. In this manner, the user may subsequently search for data to obtain a "global" overview of the toxicological data landscape. In certain circumstances, the searching of chemical structures may simply be within a database. The "similar" structures, together with the target structure, then form the initial grouping or category. If it is not possible to identify similar structures at this point, then different approaches to estimating similarity may be required.

1.4.3 Step 3 - Obtaining Toxicity Data for the Grouping or Category

Once the target and similar structures have been identified, suitable data for the structures should be retrieved and stored. At this point, the data will need to be assessed for quality and suitability (see Chapter 5). If insufficient data, or data of a low quality, are all that are available then read-across may not be possible. This may mean the use of different approaches to assess similarity or the requirement to utilise different endpoints or effects.

1.4.4 Step 4 - Definition of the Category

Combining together the target chemical, its analogues and the data will allow for the definition of the category. This will need to be examined for consistency. At this point sub-categorisation may be required, which can have the effect of reducing the number of chemicals within a category, but should strengthen the definition of the category.

1.4.5 Step 5 - Prediction of Toxicity by Read-Across

Assuming that a robust category has been obtained for the target chemical (see Chapter 6), which has been filled with suitable, high quality data, then read-across may be performed to make a prediction. This may be as simple as an interpolation of an activity or may require the bespoke development of a QSAR relating activities to the properties of the molecule.

1.4.6 Step 6 - Documentation of the Prediction

Should the read-across prediction be required for regulatory purposes, for instance, then it will need to be fully documented. This is a process by which the whole category formation and read-across approach is recorded such that, if required, it could be repeated to verify the prediction. In addition, this means that the information on which the read-across prediction is based can be assessed. All stages of the process shown in Figure 1.2 should be recorded. See Chapter 7 for more details.

1.4.6.1 Accepting or Rejecting the Prediction

The final process of read-across is to accept or reject the prediction as fit for purpose, and to attempt to determine the confidence associated with the prediction. This is intended to assess the validity of the prediction. As such it will depend on the context in which it is being used, the endpoint, the chemical substance and many other factors. For more information see Chapter 7.

1.4.7 Applying the Flow Chart Depicted in Figure 1.3

It should immediately become obvious that whilst read-across is, at first sight, a trivial activity, there are many complex steps. In theory, an experienced toxicologist and/or chemist should be able to develop their own categories and populate them with data manually. However, experience will inform the user that for many endpoints, and especially for more complex chemistry, computational techniques will be required, *e.g.* to assist in finding similar structures (using a variety of methods), storing the structures, identification of data to populate the category, performing read-across (including development of local QSARs) and recording the whole process with standardised documentation. There are now a number of computational approaches and software that can be used, these are described in Chapter 4.

To apply any predictive technique, and category formation and read-across are no exception, then a degree of understanding and expertise is required. To apply these techniques, whilst there are some excellent free tools available, expertise is required in a number of key areas.

1.5 Advantages and Disadvantages of Category Formation and Read-Across

Depending on with whom one discusses the issue, there are a variety of advantages and disadvantages of using category formation and read-across to fill data gaps for the prediction of toxicity and physico-chemical properties. Patlewicz et al.[88,89] discuss the use of read-across pragmatically; some of their opinions form the basis of the following lists.

The main advantages of category formation and read-across include:

- It is relatively cheap and rapid, at least in comparison to *in vivo* toxicity testing.
- It can form a valuable part of an Integrated Testing Strategy and can be supported by other relevant non-test data *e.g.* from *in vitro* or molecular biology testing.
- It is ethically defensible in terms of not requiring animal use.
- It promotes the philosophy of green chemistry through the efficient use of resources.
- There are numerous free softwares and databases to support grouping and read-across.
- There is considerable guidance and examples of case studies to assist the user.
- It is usually strongly based around similar mechanisms and modes of action, increasing the transparency of predictions.
- There is an increasing acceptance by regulators of predictions from read-across.
- Category formation can benefit from the advances in 21st Century Toxicology and Adverse Outcome Pathways.
- Category formation shows the limitations of existing data for making a prediction.
- It is supported by all major stakeholders including industry, regulatory and Non-Governmental Organisations.
- The development of the process of read-across and tools to achieve it have received generous support and funding in the European Union, North America and Japan.

There are, however, a number of drawbacks and disadvantages. Many specifically temper the advantages so may, at first consideration, appear contradictory.

- It does require expert use, and hence the availability of trained experts.
- Creating a category and performing read-across is at best subjective. Often different experts may reach different conclusions for the same compound.
- It requires the selection of the correct means of grouping chemicals, whether this may be based on a mechanistic, analogue or structural similarity approach.
- It requires good quality toxicity (or other) data to make a prediction.
- Confidence in a read-across estimate is reliant on a number of factors (*e.g.* robustness of the definition of the category, number of compounds in the category *etc*) which may be difficult to quantify.

1.6 Uses of Read-Across and Category Formation - Current Literature

Forming a group of chemicals and attempting read-across can, in theory, be applied to any toxicity or other type of endpoint. The report by ECHA[108] details its use to make predictions for all types of toxicities (although that does not necessarily imply acceptance by ECHA). Reports in the literature are centred around a smaller number of endpoints, typically those associated with repeat-dose toxicity, skin sensitisation, developmental toxicity and mutagenicity, with regard to human health and mammalian endpoints, and acute toxicity with regard to environmental effects. Most studies have described applications for organic chemicals (*i.e.* "classic" industrial chemicals, drugs *etc.*), whilst a smaller number have described its use to make assessments for metals and metal complexes. Further rationalisation of the use of read-across for data gap filling, with an emphasis on REACH submission, is provided by Patlewicz *et al.*[89] Table 1.3 summarises the main endpoints to which read-across has been applied at the time of writing. It is not likely to be a complete list but gives a flavour of the main endpoints and types of compounds considered.

1.7 Key Literature and Guidance for the Regulatory Use of Read-Across

In addition to the literature cited in the previous section there is much freely available guidance to the use of grouping and read-across (Patlewicz *et al.*).[88,89] Some of the key documents are summarised in Table 1.4.[5,6,110–116] It should be noted that the most comprehensive of these documents[112,113] were prepared almost simultaneously with this book. This book has not attempted to

Table 1.3 Summary of the recent literature describing grouping, category formation and read-across to predict a variety of toxicity endpoints.

Endpoint	Compounds and References
Acute rodent toxicity	Miscellaneous organic compounds;[9,10] Non-reactive compounds;[11] Nickel[12,13]
Repeat dose (haemolytic effects)	Ethylene glycol alkyl ethers[14]
Repeat dose	Ferrochromium;[15] nitrobenzenes[16]
Hepatotoxicity	Miscellaneous organic compounds[17]
Teratogenicity	Miscellaneous organic compounds[18]
Reproductive effects	Phthalates[19]
DNA binding/mutagenicity	Miscellaneous organic compounds[3,20–22]
In vivo micronucleus assay	Miscellaneous organic compounds[23]
Skin sensitisation	Miscellaneous organic compounds;[24–28] Michael acceptor electrophiles;[29–32] Compounds acting as aromatic nucleophilic substituents;[33] Alpha, beta-unsaturated carbonyls;[34] Rosin-based substances;[35] Phenyl glycidyl ethers;[36] Osmolytic prodrugs;[37] Benzoquinone[38]
Respiratory sensitisation	Miscellaneous organic compounds[39–41]
Miscellaneous mammalian effects	Long-chain aliphatic alcohols[42–43]
Acute environmental toxicity	Miscellaneous organic compounds;[44,45] S_N2 electrophiles[46]
Chronic environmental toxicity	Pharmaceuticals;[47] Fragrances[48]
Reproductive effects in fish	Dutasteride[49]
Environmental effects, fate and properties	Long-chain aliphatic alcohols[50]
Various environmental effects	Pharmaceuticals;[51] Long-chain aliphatic alcohols;[52] Veterinary products;[53] Long-chain alcohols;[54] Persistent transformation (degradation) products;[55,56] Glycol ether alkoxy acids and 1,2,4-triazole fungicides;[57] Anionic surfactants[58]
General regulatory use and guidance	There are a large number of papers that provide basic information on developing categories of relevance for regulatory use.[59–89]
Tools and approaches	Toxmatch software;[90,91] Atom-centred fragments to define groups;[92] PETROTOX software;[93] Metabolomics as a tool for grouping;[94] HESS[95]

replicate these publications but to supplement them with meaningful information, not necessarily derived from an industry basis.

1.8 Aims of this Volume

This volume introduces the reader to a rapidly growing technique in the computational, or *in silico*, prediction of toxicity, namely the creation of

Table 1.4 Summary of the key documents describing grouping, read-across and category formation.

Institution	General Information	Content	Reference
European Chemicals Agency (ECHA)	Two documents are provided (11 and 19 pages respectively) prepared by ECHA. Freely available from http://echa.europa.eu/en/support/grouping-of-substances-and-read-across	Part 1 is a brief introduction to read-across and the general considerations. Part 2 contains a very useful illustrative example for a hypothetical substance. It is anticipated that additional examples will be added to Part 2 in the future.	ECHA[110,111]
European Centre for Ecotoxicology and Toxicology of Chemicals (ECETOC)	A 192 page document, written by an industry-led task force. Freely available from http://www.ecetoc.org/technical-reports	Excellent general background on definitions, resources, workflows and endpoint specific considerations and case studies	ECETOC[112]
Organisation for Economic Cooperation and Development (OECD)	A 176 page Workshop Report prepared by the OECD. Freely available from http://www.oecd.org/env/ehs/risk-assessment/guidancedocumentsandreportsrelatedtoqsars.htm	Provides conceptual background and examples for the use of Adverse Outcome Pathways to support robust category development, and hence read-across.	OECD[113]
European Chemicals Agency (ECHA)	A 34 page document, prepared by ECHA. Freely available from http://echa.europa.eu/en/support/grouping-of-substances-and-read-across	Provides a useful introduction into how to report read-across results in IUCLID5.	ECHA[114]
European Chemicals Agency (ECHA)	A 134 page document, prepared by ECHA. Freely available from http://echa.europa.eu/documents/10162/13632/information_requirements_r6_en.pdf	Describes both with QSAR and grouping/read-across. It provides fundamental information on grouping, as well as the regulatory acceptance of predictions from (Q)SARs.	ECHA[6]
Organisation for Economic Cooperation and Development (OECD)	A 99 page document prepared by the OECD. Freely available from http://www.oecd.org/env/ehs/risk-assessment/guidancedocumentsandreportsrelatedtoqsars.htm	This develops the information provided by the JRC (see below) to define the category concept, read-across, data gap filling and reporting formats. It is supported with examples.	OECD[5]

Table 1.4 (Continued)

Institution	General Information	Content	Reference
European Commission's Joint Research Centre (EC JRC)	A 132 page document prepared by the JRC. Freely available from http://ihcp.jrc.ec.europa.eu/our_labs/predictive_toxicology/publications/articles	Background on the use of computational approaches for grouping and read-across	Worth et al[115]
European Commission's Joint Research Centre (EC JRC)	A 180 page document prepared by members of a REACH Implementation Panel (RIP). Freely available from http://ihcp.jrc.ec.europa.eu/our_labs/predictive_toxicology/publications/articles	A large number of case studies of how and where read-across has been applied.	Worth and Patlewicz[116]

categories and read-across predictions for toxicity. This volume will outline the process, methods and interpretation of results in this area. Specifically it informs the reader how to obtain toxicity data and where to go to find them, as well as how to determine the quality of such data. Techniques to find similar chemicals are discussed, with a particular emphasis on the application of Adverse Outcome Pathways (AOPs). Furthermore, it will demonstrate how these predictions can be used to meet regulatory requirements. The aims of the book are supported by a number of case studies and examples in this rapidly growing area.

1.9 Conclusions

Information on the harmful effects of chemicals to mankind and the environment is required. This has traditionally been derived from animal tests but there is an increasing desire, due to reasons of cost, ethics and regulatory pressure, to move to non-animal alternatives. Key amongst the alternatives is the application of *in silico* or computational approaches. The simplest of these *in silico* approaches is the ability to group compounds together, or form a category, and then read-across toxicity. There is a well-defined and flexible process to undertake read-across supported by many case studies and key guidance documents. There is likely to be increased use of grouping and read-across to provide predictions which are acceptable for regulatory purposes, especially for effects following repeated dose exposure.

Acknowledgement

Funding from the European Community's Seventh Framework Program (FP7/2007–2013) COSMOS Project under grant agreement no. 266835 and from Cosmetics Europe is gratefully acknowledged. Funding from the European Chemicals Agency (Service Contract No. ECHA/2008/20/ECA.203) is also gratefully acknowledged.

References

1. C. J. van Leeuwen and T. G. Vermeire, *Risk Assessment of Chemicals: An Introduction, 2nd Ed.*, Springer, Dordrecht, The Netherlands, 2007.
2. D. Krewski, D. Acosta, M. Andersen, H. Anderson, J. C. Bailar, K. Boekelheide, R. Brent, G. Charnley, V. G. Cheung, S. Green, K. T. Kelsey, N. I. Kerkvliet, A. A. Li, L. McCray, O. Meyer, R. D. Patterson, W. Pennie, R. A. Scala, G. M. Solomon, M. Stephens, J. Yager and L. Zeise, Toxicity testing in the 21st Century: A vision and strategy, *J. Toxicol. Environ. Health - Part B – Crit. Rev.*, 2010, **13**, 51.
3. S. J. Enoch and M. T. D. Cronin, A review of the electrophilic reaction chemistry involved in covalent DNA binding, *Crit. Rev. Toxicol.*, 2010, **40**, 728.

4. C. M. Ellison, M. T. D. Cronin, J. C. Madden and T. W. Schultz, Definition of the structural domain of the baseline non-polar narcosis model for *Tetrahymena pyriformis*, *SAR QSAR Environ. Res.*, 2008, **19**, 751.
5. Organisation for Economic Co-operation and Development (OECD), Environment Health and Safety Publications, Series on Testing and Assessment No. 80: *Guidance on Grouping of Chemicals*, OECD, Paris, 2007.
6. European Chemicals Agency (ECHA), *Guidance on Information Requirements and Chemical Safety Assessment, Chapter R.6: QSARs and Grouping of Chemicals*, ECHA, Helsinki, 2008.
7. M. T. D. Cronin and D. J. Livingstone, *Predicting Chemical Toxicity and Fate*, CRC Press, Boca Raton, USA, 2004.
8. M. T. D. Cronin and J. C. Madden, *In Silico Toxicology. Principles and Applications*, RSC Publishing, Cambridge, 2010.
9. O. A. Raevsky, V. Ju. Grigor'ev, E. A. Modina and A. P. Worth, Prediction of acute toxicity to mice by the Arithmetic Mean Toxicity (AMT) modelling approach, *SAR QSAR Environ. Res.*, 2010, **21**, 265.
10. O. A. Raevsky, V. Y. Grigor'ev, E. A. Liplavskaya and A. P. Worth, Prediction of acute rodent toxicity on the basis of Chemical Structure and Physicochemical Similarity, *Mol. Inf.*, 2011, **30**, 267.
11. Y. K. Koleva, M. T. D. Cronin, J. C. Madden and J. A. H. Schwöbel, Modelling acute oral mammalian toxicity. 1. Definition of a quantifiable baseline effect, *Toxicol. In Vitro*, 2011, **25**, 1281.
12. R. G. Henderson, D. Cappellini, S. K. Seilkop, H. K. Bates and A. R. Oller, Oral bioaccessibility testing and read-across hazard assessment of nickel compounds, *Regul. Toxicol. Pharmacol.*, 2012, **63**, 20.
13. R. G. Henderson, J. Durando, A. R. Oller, D. J. Merkel, P. A. Marone and H. K. Bates, Acute oral toxicity of nickel compounds, *Regul. Toxicol. Pharmacol.*, 2012, **62**, 425.
14. T. Yamada, Y. Tanaka, H. Q. Zhang, R. Hasegawa, Y. Sakuratani, O. Mekenyan, Y. Yamazoe, J. Yamada and M. Hayashi, A category approach to predicting the hemolytic effects of ethylene glycol alkyl ethers in repeated-dose toxicity, *J. Toxicol. Sci.*, 2012, **37**, 503.
15. H. Stockmann-Juvala, A. Zitting, G. Darrie, I. O. Wallinder and T. Santonen, Read-across approach in the risk assessment Ferrochromium. Case: Repeated dose toxicity, *Toxicol. Lett.*, 2009, **189**, S245.
16. Y. Sakuratani, H. Q. Zhang, S. Nishikawa, K. Yamazaki, T. Yamada, J. Yamada and M. Hayashi, Categorization of nitrobenzenes for repeated dose toxicity based on adverse outcome pathways, *SAR QSAR Environ. Res.*, 2013, **24**, 35.
17. C. M. Ellison, S. J. Enoch and M. T. D. Cronin, A review of the use of *in silico* methods to predict the chemistry of molecular initiating events related to drug toxicity, *Exp. Opin. Drug Metab. Toxicol.*, 2011, **7**, 1481.
18. S. J. Enoch, M. T. D. Cronin, J. C. Madden and M. Hewitt, Formation of structural categories to allow for read-across for teratogenicity, *QSAR Comb. Sci.*, 2009, **28**, 696.

19. E. Fabjan, E. Hulzebos, W. Mennes and A. H. Piersma, A category approach for reproductive effects of phthalates, *Crit. Rev. Toxicol.*, 2006, **36**, 695.
20. S. J. Enoch and M. T. D. Cronin, Defining the chemistry associated with DNA binding to allow for prediction of mutagenicity by grouping and read across, *Mutagenesis*, 2012, **27**, 789.
21. S. J. Enoch and M. T. D. Cronin, Development of new structural alerts suitable for chemical category formation for assigning covalent and non-covalent mechanisms relevant to DNA binding, *Mut. Res. Genet. Toxicol. Environ. Mutagen.*, 2012, **743**, 10.
22. S. J. Enoch, M. T. D. Cronin and C. M. Ellison, The use of a chemistry-based profiler for covalent DNA binding in the development of chemical categories for read-across for genotoxicity, *Altern. Lab. Anim. ATLA*, 2011, **39**, 131.
23. R. Benigni, C. Bossa and A. Worth, Structural analysis and predictive value of the rodent *in vivo* micronucleus assay results, *Mutagenesis*, 2010, **25**, 335.
24. D. W. Roberts, G. Patlewicz, P. S. Kern, F. Gerberick, I. Kimber, R. J. Dearman, C. A. Ryan, D. A. Basketter and A. O. Aptula, Mechanistic applicability domain classification of a local lymph node assay dataset for skin sensitization, *Chem. Res. Toxicol.*, 2007, **20**, 1019.
25. D. W. Roberts and A. O. Aptula, Determinants of skin sensitisation potential, *J. Appl. Toxicol.*, 2008, **28**, 377.
26. D. W. Roberts, A. O. Aptula, G. Patlewicz and C. Pease, Chemical reactivity indices and mechanism-based read-across for non-animal based assessment of skin sensitisation potential, *J. Appl. Toxicol.*, 2008, **28**, 443.
27. S. J. Enoch, J. C. Madden and M. T. D. Cronin, Identification of mechanisms of toxic action for skin sensitisation using a SMARTS pattern based approach, *SAR QSAR Environ. Res.*, 2008, **19**, 555.
28. S. J. Enoch, C. M. Ellison, T. W. Schultz and M. T. D. Cronin, A review of the electrophilic reaction chemistry involved in covalent protein binding relevant to toxicity, *Crit. Rev. Toxicol.*, 2011, **41**, 783.
29. D. W. Roberts and A. Natsch, High throughput kinetic profiling approach for covalent binding to peptides: Application to skin sensitization potency of Michael acceptor electrophiles, *Chem. Res. Toxicol.*, 2009, **22**, 592.
30. T. W. Schultz, K. Rogers and A. O. Aptula, Read-across to rank skin sensitization potential: subcategories for the Michael acceptor domain, *Contact Dermat.*, 2009, **60**, 21.
31. S. J. Enoch, M. T. D. Cronin, T. W. Schultz and J. C. Madden, Quantitative and mechanistic read across for predicting the skin sensitization potential of alkenes acting via Michael addition, *Chem. Res. Toxicol.*, 2008, **21**, 513.
32. S. J. Enoch and D. W. Roberts, Predicting skin sensitization potency for Michael acceptors in the LLNA using quantum mechanics calculations, *Chem. Res. Toxicol.*, 2013, **26**, 767.

33. S. J. Enoch, T. W. Schultz and M. T. D. Cronin, The definition of the applicability domain relevant to skin sensitization for the aromatic nucleophilic substitution mechanism, *SAR QSAR Environ. Res.*, 2012, **23**, 649.
34. Y. K. Koleva, J. C. Madden and M. T. D. Cronin, Formation of categories from structure-activity relationships to allow read-across for risk assessment: toxicity of alpha,beta-unsaturated carbonyl compounds, *Chem. Res. Toxicol.*, 2008, **21**, 2300.
35. H. P. A. Illing, T. Malmfors and L. Rodenburg, Skin sensitization and possible groupings for 'read across' for rosin based substances, *Regul. Toxicol. Pharmacol.*, 2009, **54**, 234.
36. T. Delaine, I. B. Niklasson, R. Emter, K. Luthman, A. T. Karlberg and A. Natsch, Structure-activity relationship between the in vivo skin sensitizing potency of analogues of phenyl glycidyl ether and the induction of Nrf2-dependent luciferase activity in the KeratinoSens in vitro assay, *Chem. Res. Toxicol.*, 2011, **24**, 1312.
37. J. Scheel and D. Keller, Investigation of the skin sensitizing properties of 5 osmolytic prodrugs in a Weight-of-Evidence assessment, employing *in silico*, *in vivo*, and read across analyses, *Int. J. Toxicol.*, 2012, **31**, 358.
38. D. W. Roberts and A. O. Aptula, Does the extreme skin sensitization potency of benzoquinone result from special chemistry?, *Contact Dermat.*, 2009, **61**, 332.
39. S. J. Enoch, D. W. Roberts and M. T. D. Cronin, Electrophilic reaction chemistry of low molecular weight respiratory sensitizers, *Chem. Res. Toxicol.*, 2009, **22**, 1447.
40. S. J. Enoch, D. W. Roberts and M. T. D. Cronin, Mechanistic category formation for the prediction of respiratory sensitization, *Chem. Res. Toxicol.*, 2010, **23**, 1547.
41. S. J. Enoch, M. J. Seed, D. W. Roberts, M. T. D. Cronin, S. J. Stocks and R. M. Agius, Development of mechanism-based structural alerts for respiratory sensitization hazard identification, *Chem. Res. Toxicol.*, 2012, **25**, 2490.
42. H. Sanderson, S. E. Belanger, P. R. Fisk, C. Schaefers, G. Veenstra, A. M. Nielsen, Y. Kasai, A. Willing, S. D. Dyer, K. Stanton and R. Sedlak, An overview of hazard and risk assessment of the OECD high production volume chemical category - Long chain alcohols [C-6-C-22] (LCOH), *Ecotox. Environ. Saf.*, 2009, **72**, 973.
43. G. Veenstra, C. Webb, H. Sanderson, S. E. Belanger, P. Fisk, A. Nielsen, Y. Kasai, A. Willing, S. Dyer, D. Penney, H. Certa, K. Stanton and R. Sedlak, Human health risk assessment of long chain alcohols, *Ecotox. Environ. Saf.*, 2009, **72**, 1016.
44. G. Schüürmann, R. U. Ebert and R. Kühne, Quantitative read-across for predicting the acute fish toxicity of organic compounds, *Environ. Sci. Technol.*, 2011, **45**, 4616.

45. R. Kühne, R.-U. Ebert, P. C. von der Ohe, N. Ulrich, W. Brack and G. Schüürmann, Read-across prediction of the acute toxicity of organic compounds toward the water flea *Daphnia magna*, *Mol. Inf.*, 2013, **32**, 108.
46. D. W. Roberts, T. W. Schultz, E. M. Wolf and A. O. Aptula, Experimental reactivity parameters for toxicity modeling: application to the acute aquatic toxicity of S(N)2 electrophiles to *Tetrahymena pyriformis*, *Chem. Res. Toxicol.*, 2010, **23**, 228.
47. J. P. Berninger and B. W. Brooks, Leveraging mammalian pharmaceutical toxicology and pharmacology data to predict chronic fish responses to pharmaceuticals, *Toxicol. Lett.*, 2010, **193**, 69.
48. E. Rorije, T. Aldenburg and W. Peijnenburg, Read-across estimates of aquatic toxicity for selected fragrances, *Altern. Lab. Anim. ALTA*, 2013, **41**, 77.
49. L. Margiotta-Casaluci, R. E. Hannah and J. P. Sumpter, Mode of action of human pharmaceuticals in fish: The effects of the 5-alpha-reductase inhibitor, dutasteride, on reproduction as a case study, *Aquat. Toxicol.*, 2013, **128–129**, 113.
50. P. R. Fisk, R. J. Wildey, A. E. Girling, H. Sanderson, S. E. Belanger, G. Veenstra, A. Nielsen, Y. Kasai, A. Willing, S. D. Dyer and K. Stanton, Environmental properties of long chain alcohols. Part 1: Physicochemical, environmental fate and acute aquatic toxicity properties, *Ecotox. Environ. Saf.*, 2009, **72**, 980.
51. K. Olejniczak, R. T. Williams, T. Kuehler and P. Spindler, Summary of workshop on use of data developed during drug development for read across to environmental analyses, *Drug Disc. J.*, 2007, **41**, 201.
52. S. E. Belanger, H. Sanderson, P. R. Fisk, C. Schaefers, S. M. Mudge, A. Willing, Y. Kasai, A. M. Nielsen, S. D. Dyer and R. Toy, Assessment of the environmental risk of long-chain aliphatic alcohols, *Ecotox. Environ. Saf.*, 2009, **72**, 1006.
53. B. W. Brooks, R. A. Brain, D. B. Huggett and G. T. Ankley, Risk assessment considerations for veterinary medicines in aquatic ecosystems, *Veterin. Pharmac. Environ.*, 2009, **1018**, 205.
54. C. Schaefers, U. Boshof, H. Juerling, S. E. Belanger, H. Sanderson, S. D. Dyer, A. M. Nielsen, A. Willing, K. Gamon, Y. Kasai, C. V. Eadsforth, P. R. Fisk and A. E. Girling, Environmental properties of long-chain alcohols, Part 2: Structure-activity relationship for chronic aquatic toxicity of long-chain alcohols, *Ecotox. Environ. Saf.*, 2009, **72**, 996.
55. C. J. Sinclair and A. B. A. Boxall, Ecotoxicity of Transformation Products, *Transformation Products of Synthetic Chemicals in the Environment*, 2009, **2**, 177.
56. B. I. Escher and K. Fenner, Recent advances in environmental risk assessment of transformation products, *Environ. Sci. Technol.*, 2011, **45**, 3835.
57. S. A. B. Hermsen, T. E. Pronka, E.-J. van den Brandhof, L. T. M. van der Ven and A. H. Piersma, Chemical class-specific gene expression

changes in the zebrafish embryo after exposure to glycol ether alkoxy acids and 1,2,4-triazole antifungals, *Reprod. Toxicol.*, 2011, **32**, 245.
58. G. Koennecker, J. Regelmann, S. Belanger, K. Gamon and R. Sedlak, Environmental properties and aquatic hazard assessment of anionic surfactants: Physico-chemical, environmental fate and ecotoxicity properties, *Ecotox. Environ. Saf.*, 2011, **74**, 1445.
59. J. Ahlers, F. Stock and B. Werschkun, Integrated testing and intelligent assessment-new challenges under REACH, *Environ. Sci. Pollut. Res.*, 2008, **15**, 565.
60. A. Bassan and A. P. Worth, The integrated use of models for the properties and effects of chemicals by means of a structured workflow, *QSAR Comb. Sci.*, 2008, **27**, 6.
61. H. Greim, Evidence-based toxicological evaluation of chemicals by group summaries, *Bundesdesundheitsblatt-Gesundheitsforschung-Gesundheitsschutz*, 2008, **51**, 1417.
62. U. Lahl and U. Gundert-Remy, The use of (Q)SAR methods in the context of REACH, *Toxicol. Mech. Meth.*, 2008, **18**, 149.
63. W. Lilienblum, W. Dekant, H. Foth, T. Gebel, J. G. Hengstler, R. Kahl, P. -J. Kramer, H. Schweinfurth and K. -M. Wollin, Alternative methods to safety studies in experimental animals: role in the risk assessment of chemicals under the new European Chemicals Legislation (REACH), *Arch. Toxicol.*, 2008, **82**, 211.
64. H. Niska, K. Tuppurainen, J. -P. Skoen, A. K. Mallett and M. Kolehmainen, Characterisation of the chemical and biological properties of molecules with QSAR/QSPR and chemical grouping, and its application to a group of alkyl ethers, *SAR QSAR Environ. Res.*, 2008, **19**, 263.
65. F. L. Assem and L. S. Levy, A review of current toxicological concerns on vanadium pentoxide and other vanadium compounds: gaps in knowledge and directions for future research, *J. Toxicol. Environ. Heal. B. Crit. Rev.*, 2009, **12**, 289.
66. G. Schaafsma, E. D. Kroese, E. L. J. P. Tielemans, J. J. M. van de Sandt and C. J. van Leeuwen, REACH, non-testing approaches and the urgent need for a change in mind set, *Regul. Toxicol. Pharmacol.*, 2009, **53**, 70.
67. M. Stenberg, A. Linusson, M. Tysklind and P. L. Andersson, A multivariate chemical map of industrial chemicals – Assessment of various protocols for identification of chemicals of potential concern, *Chemosphere*, 2009, **76**, 878.
68. K. van Leeuwen, T. W. Schultz, T. Henry, B. Diderich and G. D. Veith, Using chemical categories to fill data gaps in hazard assessment, *SAR QSAR Environ. Res.*, 2009, **20**, 207.
69. J. A. Vonk, R. Benigni, M. Hewitt, M. Nendza, H. Segner, D. van de Meent and M. T. D. Cronin, The use of mechanisms and modes of toxic action in Integrated Testing Strategies: The report and recommendations

of a Workshop held as part of the European Union OSIRIS Integrated Project, *Altern. Lab. Anim. ATLA*, 2009, **37**, 557.
70. M. Hewitt, C. M. Ellison, S. J. Enoch, J. C. Madden and M. T. D. Cronin, Integrating (Q)SAR models, expert systems and read-across approaches for the prediction of developmental toxicity, *Reprod. Toxicol.*, 2010, **30**, 147.
71. E. Hulzebos and I. Gerner, Weight factors in an Integrated Testing Strategy using adjusted OECD principles for (Q)SARs and extended Klimisch codes to decide on skin irritation classification, *Regul. Toxicol. Pharmacol.*, 2010, **58**, 131.
72. E. Hulzebos, S. Gunnarsdottir, J.-P. Rila, Z. C. Dang and E. Rorije, An Integrated Assessment Scheme for assessing the adequacy of (eco)toxicological data under REACH, *Toxicol. Lett.*, 2010, **198**, 255.
73. S. R. Vink, J. Mikkers, T. Bouwman, H. Marquart and E. D. Kroese, Use of read-across and tiered exposure assessment in risk assessment under REACH – A case study on a phase-in substance, *Regul. Toxicol. Pharmacol.*, 2010, **58**, 64.
74. M. J. Winter, S. F. Owen, R. Murray-Smith, G. H. Panter, M. J. Hetheridge and L. B. Kinter, Using data from drug discovery and development to aid the aquatic environmental risk assessment of human pharmaceuticals: concepts, considerations, and challenges, *Integrated Environmental Assessment and Management*, 2010, **6**, 38.
75. S. Wu, K. Blackburn, J. Amburgey, J. Jaworska and T. Federle, A framework for using structural, reactivity, metabolic and physicochemical similarity to evaluate the suitability of analogs for SAR-based toxicological assessments, *Regul. Toxicol. Pharmacol.*, 2010, **56**, 67.
76. S. Adler, D. Basketter, S. Creton, O. Pelkonen, J. van Benthem, V. Zuang, K. E. Andersen, A. Angers-Loustau, A. Aptula, A. Bal-Price, E. Benfenati, U. Bernauer, J. Bessems, F. Y. Bois, A. Boobis, E. Brandon, S. Bremer, T. Broschard, S. Casati, S. Coecke, R. Corvi, M. Cronin, G. Daston, W. Dekant, S. Felter, E. Grignard, U. Gundert-Remy, T. Heinonen, I. Kimber, J. Kleinjans, H. Komulainen, R. Kreiling, J. Kreysa, S. B. Leite, G. Loizou, G. Maxwell, P. Mazzatorta, S. Munn, S. Pfuhler, P. Phrakonkham, A. Piersma, A. Poth, P. Prieto, G. Repetto, V. Rogiers, G. Schoeters, M. Schwarz, R. Serafimova, H. Tahti, E. Testai, J. van Delft, H. van Loveren, M. Vinken, A. Worth and J.-M. Zaldivar, Alternative (non-animal) methods for cosmetics testing: current status and future prospects - 2010, *Arch. Toxicol.*, 2011, **85**, 367.
77. E. Antignac, G. J. Nohynek, T. Re, J. Clouzeau and H. Toutain, Safety of botanical ingredients in personal care products/cosmetics, *Fd Chem. Toxicol.*, 2011, **49**, 324.
78. K. Blackburn, D. Bjerke, G. Daston, S. Felter, C. Mahony, J. Naciff, S. Robison and S. Wu, Case studies to test: A framework for using structural, reactivity, metabolic and physicochemical similarity to

evaluate the suitability of analogs for SAR-based toxicological assessments, *Regul. Toxicol. Pharmacol.*, 2011, **60**, 120.
79. N. Carmichael, M. Bausen, A. R. Boobis, S. M. Cohen, M. Embry, C. Fruijtier-Poelloth, H. Greim, R. Lewis, M. E. Meek, H. Mellor, C. Vickers and J. Doe, Using mode of action information to improve regulatory decision-making: An ECETOC/ILSI RF/HESI workshop overview, *Crit. Rev. Toxicol.*, 2011, **41**, 175.
80. M. T. D. Cronin, S. J. Enoch, M. Hewitt and J. C. Madden, formation of Mechanistic Categories and Local Models to Facilitate the Prediction of Toxicity, *ALTEX - Altern. Anim. Exper.*, 2011, **28**, 45.
81. G. Patlewicz, M. W. Chen, C. A. Bellin, Non-testing approaches under REACH - help or hindrance? Perspectives from a practitioner within industry, *SAR QSAR Environ. Res.*, 2011, **22**, 67.
82. P. C. von der Ohe, V. Dulio, J. Slobodnik, E. De Deckere, R. Kühne, R.-U. Ebert, A. Ginebreda, W. De Cooman, G. Schüürmann and W. Brack, A new risk assessment approach for the prioritization of 500 classical and emerging organic microcontaminants as potential river basin specific pollutants under the European Water Framework Directive, *Sci. Tot. Environ.*, 2011, **409**, 2064.
83. P. L. Bishop, J. R. Manuppello, C. E. Willett and J. T. Sandler, Animal use and lessons learned in the U. S. High Production Volume Chemicals Challenge Program, *Environ. Heal. Persp.*, 2012, **120**, 1631.
84. O. Deeb and M. Goodarzi, In silico quantitative structure toxicity relationship of chemical compounds: some case studies, *Curr. Drug Saf.*, 2012, **7**, 289.
85. S. Modi, M. Hughes, A. Garrow and A. White, The value of *in silico* chemistry in the safety assessment of chemicals in the consumer goods and pharmaceutical industries, *Drug Disc. Today*, 2012, **17**, 135.
86. F. A. Quintero, S. J. Patel, F. Munoz and M. S. Mannan, Review of existing QSAR/QSPR models developed for properties used in hazardous chemicals classification system, *Indust. Engin. Chem. Res.*, 2012, **51**, 16101.
87. G. Y. Patlewicz and D. R. Lander, A step change towards risk assessment in the 21st century, *Front. Biosci.*, 2013, **5**, 418.
88. G. Patlewicz, D. W. Roberts, A. Aptula, K. Blackburn and B. Hubesch, Workshop: Use of "read-across" for chemical safety assessment under REACH, *Regul. Toxicol. Pharmacol.*, 2013, **65**, 226.
89. G. Patlewicz, N. Ball, E. D. Booth, E. Hulzebos, E. Zvinavashe and C. Hennes, Use of category approaches, read-across and (Q)SAR: General considerations, *Regul. Toxicol. Pharmacol.* 2013 In press.
90. A. Gallegos-Saliner, A. Poater, N. Jeliazkova, G. Patlewicz and A. P. Worth, Toxmatch-A chemical classification and activity prediction tool based on similarity measures, *Regul. Toxicol. Pharmacol.*, 2008, **52**, 77.

91. G. Patlewicz, N. Jeliazkova, A. Saliner-Gallegos and A. P. Worth, Toxmatch - a new software tool to aid in the development and evaluation of chemically similar groups, *SAR QSAR Environ. Res.*, 2008, **19**, 397.
92. R. Kühne, R.-U. Ebert and G. Schüürmann, Chemical domain of QSAR models from atom-centered fragments, *J. Chem. Inf. Mod.*, 2009, **49**, 2660.
93. A. D. Redman, T. F. Parkerton, J. A. McGrath and D. M. Di Toro, PETROTOX: An aquatic toxicity model for petroleum substances, *Environ. Toxicol. Chem.*, 2012, **31**, 2498.
94. B. van Ravenzwaay, M. Herold, H. Kamp, M. D. Kapp, E. Fabian, R. Looser, G. Krennrich, W. Mellert, A. Prokoudine, V. Strauss, T. Walk, J. Wiemer, Metabolomics: A tool for early detection of toxicological effects and an opportunity for biology based grouping of chemicals-From QSAR to QBAR, *Mut. Res. Genet. Toxicol. Environ. Mutagen.*, 2012, **746**, 144.
95. Y. Sakuratani, H. Q. Zhang, S. Nishikawa, K. Yamazaki, T. Yamada, J. Yamada, K. Gerova, G. Chankov, O. Mekenyan and M. Hayashi, Hazard Evaluation Support System (HESS) for predicting repeated dose toxicity using toxicological categories. *SAR QSAR Environ. Res.*, **24**, 5, 617.
96. A. Lacassagne, N. P. Buu-Hoi, R. Daudel and F. Zajdela, The relation between carcinogenic activity and the physical and chemical properties of angular benzacridines, *Advanc. Cancer Res.*, 1956, **4**, 315.
97. D. M. Sanderson and E. G. Earnshaw, Computer-prediction of possible toxic action from chemical structure – the DEREK system, *Hum. Exp. Toxicol.*, 1991, **10**, 261.
98. R. G. Clements, J. V. Nabholz, M. G. Zeeman and C. M. Auer, The application of structure-activity relationships (SARs) in the aquatic toxicity evaluation of discrete organic chemicals. *SAR QSAR Environ. Res.*, 1995, **3**, 203.
99. Y. T. Woo and D. Y. Lai, OncoLogic: A Mechanism-Based Expert System for Predicting the Carcinogenic Potential of Chemicals, in *Predictive Toxicology*, ed. C. Helma, CRC Press, Boca Raton, FL, 2005, p 385.
100. G. Dupuis and C. Benezra, *Allergic Contact Dermatitis to Simple Chemicals: A Molecular Approach,* Marcel Dekker, New York, 1982.
101. E. J. Lien, *SAR: Side Effects and Drug Design,* Marcel Dekker, New York, 1987.
102. J. Ashby and R. W. Tennant, Chemical structure, *Salmonella* mutagenicity and extent of carcinogenicity as indicators of genotoxic carcinogenesis among 222 chemicals tested in rodents by the U.S. NCI/NTP, *Mutat. Res.*, 1988, **204**, 17.
103. Commission of the European Communities, *White Paper: Strategy for a Future Chemicals Policy*, 2001. Available from: http://ec.europa.eu/environment/chemicals/reach/background/white_paper.htm
104. J. S. Jaworska, M. Comber, C. Auer and C. J. van Leeuwen, Summary of a Workshop on regulatory acceptance of (Q) SARs for human health and environmental endpoints, *Environ. Health Persp.*, 2003, **111**, 1358.

105. M. T. D. Cronin, J. D. Walker, J. S. Jaworska, M. H. I. Comber, C. D. Watts and A. P. Worth, Use of QSARs in international decision-making frameworks to predict ecologic effects and environmental fate of chemical substances, *Environ. Health Persp.*, 2003, **111**, 1376.
106. M. T. D. Cronin, J. S. Jaworska, J. D. Walker, M. H. I. Comber, C. D. Watts and A. P. Worth, Use of QSARs in international decision-making frameworks to predict health effects of chemical substances, *Environ. Health Persp.*, 2003, **111**, 1391.
107. A. P. Worth, the Role of QSAR Methodology in the Regulatory Assessment of Chemicals, in *Recent Advances in QSAR Studies: Methods and Applications*, ed. T. Puzyn, J. Leszczynski and M. T. D. Cronin, Springer, Heidelberg, Germany, 2010, p 367.
108. European Chemicals Agency (ECHA), *The Use of Alternatives to Testing on Animals for the REACH Regulation 2011*, ECHA, Helsinki, ECHA-11-R-004.2-EN, 2011.
109. H. Spielmann, U. G. Sauer and O. Mekenyan, A critical evaluation of the 2011 ECHA reports on compliance with the REACH and CLP regulations and on the use of alternatives to testing on animals for compliance with the REACH Regulation, *Altern. Lab. Anim. ATLA*, 2011, **39**, 481.
110. European Chemicals Agency (ECHA), *Grouping of Substances and Read-Across Approach. Part 1: Introductory Note*, ECHA, Helsinki, ECHA-13-R-02-EN, 2013.
111. European Chemicals Agency (ECHA), *Read-Across Illustrative Example. Part 2. Example 1 – Analogue Approach: Similarity Based on Breakdown Products*, ECHA, Helsinki, ECHA-13-R-03-EN, 2013.
112. European Centre for Ecotoxicology and Toxicology of Chemicals (ECETOC), *Category Approaches, Read-across, (Q)SAR. Technical Report No 116*. ECETOC, Brussels, Belgium, 2012.
113. Organisation for Economic Co-operation and Development (OECD), *Report of the Workshop on Using Mechanistic Information in Forming Chemical Categories*, OECD Series on Testing and Assessment No. 138, OECD, Paris, 2011.
114. European Chemicals Agency (ECHA), *Practical Guide 6: How to Report Read-Across and Categories*, ECHA, Helsinki, ECHA-10-B-11-EN, 2009.
115. A. Worth, A. Bassan, E. Fabjan, A. Gallegos-Saliner, T. Netzeva, G. Patlewicz, M. Pavan and I. Tsakovska. *The Use of Computational Methods in the Grouping and Assessment of Chemicals - Preliminary Investigations*, Office for Official Publications of the European Communities, Luxembourg, 2007.
116. A. Worth and G. Patlewicz, *A Compendium of Case Studies that Helped Shape the REACH Guidance on Chemical Categories and Read Across*, EUR 22481 EN – DG Joint Research Centre, Institute IHCP, Office for Official Publications of the European Communities, Luxembourg, 2007.

CHAPTER 2
Approaches for Grouping Chemicals into Categories

S. J. ENOCH* AND D. W. ROBERTS

Liverpool John Moores University, School of Pharmacy and Chemistry, Byrom Street, Liverpool, L3 3AF, England
*E-mail: s.j.enoch@ljmu.ac.uk

2.1 Introduction

Chemical category formation and subsequent read-across analysis have been suggested as being essential if the objectives of REACH are going to be achieved without the excessive use of animals.[1-3] The use of chemical category approaches is common in a number of regulatory environments outside of the European Union, namely in the United States and Canada. According to the Organisation for Economic Co-operation and Development (OECD) a chemical category is defined as 'a group of chemicals whose physicochemical and toxicological properties are likely to be similar or follow a regular pattern as a result of structural similarity, these structural similarities may create a predictable pattern in any or all of the following parameters: physicochemical properties, environmental fate and environmental effects, and human health effects'.[2] On a practical level this process involves treating a closely related (or similar) group of chemicals as a category. Within the category toxicological data will exist for some, but not all of the chemicals for the endpoints of interest. Thus data gaps may exist for specific chemicals for each of the endpoints of interest. It is important to realise that differing data gaps are likely to exist for differing chemicals within the category. It is for these data gaps that read-across methods can be utilised to make predictions for the missing toxicological and/or physicochemical data. To this end, the OECD in conjunction with the European Chemicals Agency[3] have funded the development of the freely available OECD QSAR Toolbox for category formation and read-across (www.qsartoolbox.org).

2.2 Methods of Defining Chemical Similarity Useful in Category Formation

The fundamental requirement for the development of a category suitable for predicting toxicological effects is the ability to group chemicals together based on a common molecular initiating event. The molecular initiating event (MIE) is the interaction between a chemical and the biological system that results in the initiation of the biological cascade leading to an adverse outcome.[4] For example, the formation of a covalent bond between a protein and an electrophilic chemical has been shown to be important in a number of toxicological endpoints such as skin and respiratory sensitisation.[5-9] It is only when the chemicals within a category all act via the same mechanism that read-across methods can be used to make predictions enabling toxicological data gaps to be filled. Given the variety of toxicological endpoints and the associated number of mechanisms of action no single method for grouping chemicals exists which can be applied universally. This is especially true if one considers that for some endpoints specific mechanistic pathways remain unknown (consider the many and complex biological pathways that can potentially be disrupted leading to reproductive abnormalities).[10] In contrast, chemical categories suitable for predicting physicochemical properties are usually based purely on structural similarity as there is no equivalent event analogous to the MIE for these endpoints. In general three approaches to category formation are possible: the mechanistic approach based on knowledge of an MIE, the analogue approach and chemoinformatics approaches. All three approaches can be used to create categories for toxicological endpoints. In contrast, only the analogue and chemoinformatic approaches are suitable for creating categories for physicochemical data.

2.3 Analogue Based Category

The simplest method for chemical category formation involves identifying a functional group within the target chemical and then selecting chemicals from a database containing the same functional group. This is the analogue approach to category formation and is typically used for simple chemicals such as a series of aliphatic aldehydes. In this approach knowledge of the underlying

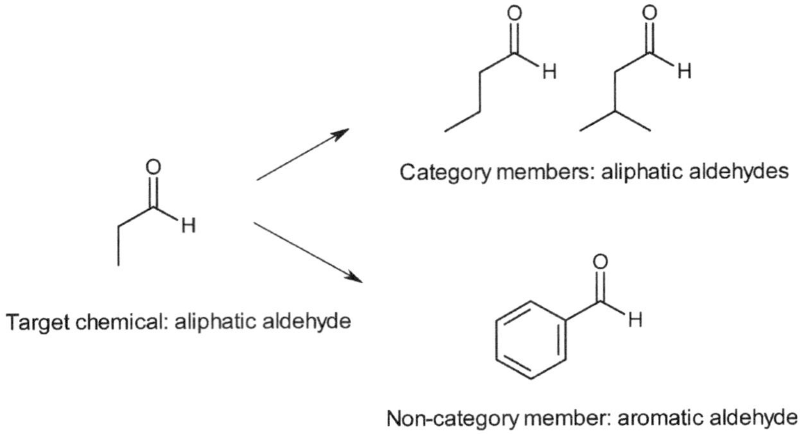

Figure 2.1 Analogue based category developed based on the presence of an aldehyde functional group in the target chemical.

mechanism is usually not known or does not exist (in the case of physicochemical data). This approach involves identifying functional groups and/or chemical elements that are present in the target chemical. Chemicals are then selected from a database containing toxicological/physicochemical data that contain only the same set of functional groups as category members. Importantly, chemicals with additional functional groups not present in the target chemical are excluded from the category. In some cases calculated parameters such as hydrophobicity are used to provide an additional measure of chemical similarity. This approach to category formation has been used recently for a number of endpoints including developmental toxicity.[11] Figure 2.1 illustrates the analogue approach for an aliphatic aldehyde target chemical showing that only other aliphatic aldehydes are included in the category (the aromatic aldehyde is excluded).

2.4 Common Mechanism of Action

Knowledge regarding the mechanism of action, especially relating to the molecular initiating event is one of the most powerful and transparent ways in which a chemical category can be formed. Category formation using mechanistic knowledge relies on the definition of structural alerts that define the key features of a molecule that are required in order for it to interact with a biological system and initiate a toxicity pathway. In general, two types of structural alerts have been developed in the literature for category formation depending on the nature of the molecular initiating event:

- Structural alerts defining the molecular features related to the formation of a covalent bond between a chemical and a biological macromolecule;

- Structural alerts defining the molecular features related to the formation of a non-covalent interaction between a chemical and a biological macromolecule. This type of structural alert covers mechanisms such as those mediated by an interaction with a biological receptor.

These mechanism-based structural alerts are used to define the chemical similarity required for category formation. In practice this is achieved by analysing the target chemical (for which a data gap exists) for the presence of one or more structural alerts related to the relevant MIE. This is known as 'profiling' the target chemical. Assuming the target chemical contains a structural alert relevant databases are then analysed for chemicals containing the same structural alert (known as 'profiling a database'). Chemicals within the database that contain only the structural alert present in the target chemical are then grouped into a mechanism-based category. The experimental data for these analogues can then be used to make a prediction for the endpoint of interest for the target chemical. Importantly, this approach cannot be used if the target chemical does not contain a structural alert related to the MIE of interest. In such cases alternative methods to define chemical similarity can be used (such as 2D similarity).

The use of structural alerts to define fragments within a molecule associated with a given MIE is well established, especially for cases where the MIE is either covalent binding to DNA or a protein. Importantly, the presence or absence of these structural alerts does not give any indication of toxicity. This is because these types of structural alerts are related to a given MIE and not a toxicity endpoint. For example, it is possible that chemical A contains a structural alert related to covalent protein binding but does not cause toxicity. In a traditional expert system this chemical might be considered as a false positive in that it contains a structural alert but is not toxic. However, consider the scenario in which a regulator wishes to fill a data gap for chemical A in which the MIE is covalent protein binding. In this case the knowledge of the organic chemistry leading to covalent protein binding can be used to create a category of mechanistically related analogues. If inspection of the available experimental data for the analogues showed them to be non-toxic then chemical A would be predicted also to be non-toxic. Unlike a traditional expert system in which the presence of an alert gives an indication of toxicity, in this example knowledge of the MIE has been used to make a prediction about the absence of toxicity. It is important to realise that the same structural alert might be used to create a second category of chemicals for a different endpoint for which covalent protein binding is also the MIE. In this category it is entirely possible that for this second endpoint chemical A may be predicted to be toxic. The explanation may be due to the relative levels of chemical reactivity required to elicit a toxicological response for the two different endpoints.

A similar argument can be made as to why the absence of a structural alert designed for category formation cannot be used as an indicator for the absence of toxicity. This is due to the fact that no matter how extensive a set of structural alerts are for a given MIE there will be areas of chemical space that have not been analysed. These unexplored areas of chemical space may contain new structural alerts. This coverage of chemical space problem applies to all types of structural alerts regardless of the MIE (*i.e.* covalent versus non-covalent) or application (structural alerts used in traditional expert systems or used in category formation). This chemical space coverage problem can be addressed by the development of experimental *in-chemico* assays that are used in an attempt to define the area of chemical space applicable to a given MIE. However, even this approach cannot guarantee complete coverage as there may be chemicals not yet synthesised that could trigger a MIE.

2.4.1 Structural Alerts for Developing Categories for Endpoints in Which Covalent Bond Formation is the Molecular Initiating Event

A number of studies have shown how structural alerts related to covalent binding to either proteins or DNA can be used to develop categories suitable for filling data gaps, using read-across, for several toxicities relevant to human health. The applicable set of structural alerts are given in parenthesis: genotoxicity[12] (covalent DNA binding), skin sensitisation[13,14] and respiratory sensitisation[15] (both covalent protein binding). Two recent review articles have outlined the mechanistic organic chemistry associated with the known structural alerts for covalent bond formation with either a protein or DNA.[7,16] This documentation of the mechanistic chemistry allows categories developed with these structural alerts to have a clear mechanistic rationale in terms of the molecular initiating event. An example of such a structural alert is shown in Figure 2.2. This alert relates to the ability of an aromatic amine to undergo metabolic activation to form an electrophilic nitrenium ion. This nitrenium ion is then capable of binding covalently to DNA.

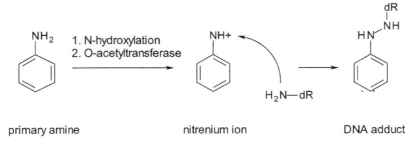

Figure 2.2 Example structural alert for an aromatic amine and the associated electrophilic reaction chemistry that explains the formation of a covalent adduct with DNA (dR = deoxyribose chain).

2.4.2 Structural Alerts for Developing Categories for Endpoints in Which a Non-Covalent Interaction is the Molecular Initiating Event

An area in which considerable research still needs to be carried out is that in relation to the development and use of structural alerts for chemicals acting via non-covalent mechanisms, especially receptor interactions. Currently, only a single profiler of this type has been developed, enabling a category to be constructed for chemicals capable of binding to the oestrogen receptor.[17] Binding to the oestrogen receptor is an important molecular initiating event for a number of toxicity endpoints including developmental toxicity.[10] This profiler uses 2D structural alerts bounded by molecular weight ranges to define the characteristics of chemicals known to bind to the oestrogen receptor based on the results of analysis of data from *in vitro* assays (Figure 2.3).

The approach of using 2D structural alerts to form categories for receptor mediated effects is not without problems. In contrast to toxicity mediated via the formation of a covalent bond (which can be easily rationalised in terms of the presence or absence of a 2D fragment within a molecule), this type of toxicity is dependent on the shape and electrostatics (including hydrogen bonding potential) of the entire molecule. This is clearly a 3D effect and thus using 2D structural alerts to define the types of chemicals able to bind to a receptor has limitations. The major one being that two molecules may not share a common 2D structural alert but might be similar in terms of their 3D shape and electrostatics. Clearly, there is scope for the application of pharmacophore methods that are able to deal with the definition of structural alerts in 3D.

2.5 Chemoinformatics

Chemoinformatics based similarity measures have also been shown to be of use in the development of chemical categories for a number of endpoints, including

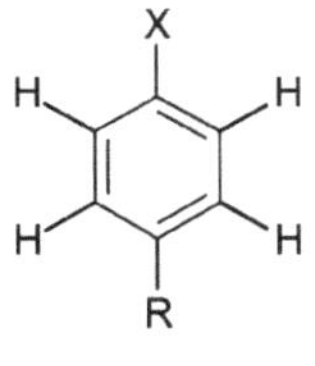

$X = OH, NH_2$

$R = alkyl$

Figure 2.3 Example of a structural alert taken from the oestrogen receptor binding profiler (available in the OECD QSAR Toolbox).

Figure 2.4 Example fingerprint for two chemicals, A and B.

skin sensitisation and developmental toxicity.[18,19] In addition, these methods have been widely used in the drug discovery paradigm for locating similar chemicals from large chemical inventories.[20,21] This makes them very useful for clustering large datasets in order to select representative chemicals from each cluster for further analysis. The primary example of this approach in the scientific literature makes use of a range so-called fingerprint methods. Such methods involve encoding the structural information within a molecule as a bit string in which each 'bit' indicates the presence (if the bit is set as a 1) or absence (if the bit is set as 0) of a particular molecular feature. Encoding a target chemical and all the chemicals in a database into such fingerprints enables them to be compared using computational measures of similarity. These measures enable the similarity between the target and each of the chemicals in the database to be assigned an integer value between 0 and 1. The closer the similarity value is to 1 the more similar the chemicals. A similarity cut-off value is then chosen that determines whether a chemical in the database is sufficiently similar to the target chemical for it to be considered as part of the category. It has been suggested that this value should be 0.6 or above.[22] The concept of bit-strings and how they can be compared to one another using the Tanimoto coefficient as the measure of similarity is summarised in Figure 2.4.

In order to assess how similar chemical A is to chemical B a similarity coefficient can be used, for example the Tanimoto coefficient (although there are numerous others, for an outline of common similarity metrics see Gillett et al.[21]). The Tanimoto coefficient requires three values; the number of 'bits' set to 1 in both fingerprints (values 'a' and 'b' in Figure 2.4) and the number of 'bits' set to 1 that both strings have in common (value 'c' in Figure 2.4). In terms of chemistry the 'bits' set to one represent the presence of functional groups. Placing these values into the formula for the Tanimoto coefficient results in a similarity value of 0.43 (the Tanimoto coefficient is as shown in Figure 2.5).

This approach to category formation has been implemented in the Toxmatch software (freely available from http://ihcp.jrc.ec.europa.eu/). This software has been used to develop categories for skin sensitisation[18] and teratogenicity[22] (see Chapter 4 for more information on tools for grouping).

$$S_{AB} = \frac{c}{a + b - c} = \frac{3}{6 + 4 - 3} = 0.43$$

Figure 2.5 Similarity between the chemical fingerprints using the Tanimoto coefficient.

2.6 The Use of Experimental Data to Support the Development of Profilers for Chemical Category Formation

Inspection of the literature (and tools such as the OECD QSAR Toolbox) shows there to be an abundance of *in vivo* toxicological data covering a wide range of endpoints. However, analysing *in vivo* data in order to develop an *in silico* profiler is fraught with difficulty due to the fact that biological systems contain multiple, complex and competing mechanisms. In addition, *in vivo* databases tend to be focussed on relatively small areas of chemical space and thus important areas of chemistry related to potential molecular initiating events may be absent from *in silico* profilers.

Despite these limitations in the use of *in vivo* data a number of studies have demonstrated that mechanism based structural alerts can be defined for certain endpoints such as skin sensitisation[6,23–26], respiratory sensitisation[8,15] and acute aquatic toxicity.[27–31] These structural alerts have been used in the development of *in silico* based profilers for these endpoints (and others) in which the formation of a covalent adduct is the molecular initiating event.[7] It is important to realise that for these endpoints a single event, the formation of a covalent adduct between a protein and an exogenous chemical, dominates the biological pathway and hence toxicity. This is especially evident in skin sensitisation[13,24,26,32–38] and acute aquatic toxicity[39–41] where chemicals with a common electrophilic mechanism can be well modelled using molecular descriptors related to chemical reactivity (or electrophilicity).

The use of *in vitro* data typically has an advantage over *in vivo* data in the development of *in silico* based profilers as it tends to be focussed on a single, key step in the biological pathway, usually the molecular initiating event. A recent study used the wide variety of structural alerts derived from the Ames assay to derive *in silico* profilers for chemicals able to form covalent adducts with DNA.[12,16,42] A number of historical and more recent research efforts have utilised experimental chemistry to explore the structural domain of a given molecular initiating event in a (relatively) simple and systematic manner.[43–45] This type of approach recently has been given the term *in chemico* data to distinguish it from *in vitro* and *in vivo* data.[24,43] This approach has been most widely used to support the development of structural alerts for covalent binding to proteins.[26,46,47] In addition, *in chemico* data also allows chemicals to be sub-categorised by rate of the reaction between a model nucleophile and a chemical. These type of data typically come from historical literature sources (for a detailed review see Schwöbel *et al.*[48]) or *in chemico* assays.[26,35,41,43,46,49–51] Several studies have shown how such potency information can be used to rationalise the skin sensitisation potential of chemicals assigned to mechanism-based categories.[26,38]

2.7 Adverse Outcome Pathways

All *in silico* profilers discussed in this chapter are developed on the basis that they are grouping chemicals together based on a common MIE. This is the key initial interaction between a chemical and a biological system that initiates the biological cascade that leads to a toxic outcome. For example, for a number of endpoints the MIE is the formation of a covalent bond with a protein. If this step does not occur then the chemical cannot initiate the biological cascade that leads to toxicity. As outlined in Chapter 3 the biological cascade leading to a toxic outcome has been defined as an adverse outcome pathway (AOP).[4,52] A key advantage of developing an AOP for a given endpoint is that it enables the definition of additional key events that may contribute to the toxic outcome. If *in vitro* assays can be developed that can model these additional key events then further *in silico* profilers could be developed. These sub-profilers could be used after an initial profiler to enable the development of a category containing chemicals that are as mechanistically similar as possible, based on all of the key events in the AOP, not just the MIE. This approach would be a clear advantage to the current category approach (as implemented in tools such as the OECD QSAR Toolbox) in which only knowledge of the molecular initiating event is used as the mechanistic basis for category formation. This process of profiling and sub-profiling using a combination of a MIE based profiler and several key event sub-profilers data is summarised in Figure 2.6.

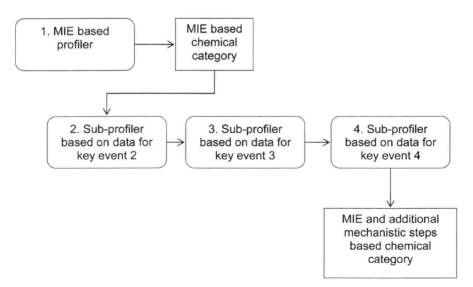

Figure 2.6 Scheme to illustrate the use of profilers and sub-profilers to develop mechanistically similar categories using the AOP concept.

The adverse outcome pathway approach to category formation is currently in its infancy, with the current focus being on skin sensitisation and toxicity to the liver. However, the concept of using an AOP to guide *in vitro* testing enabling *in silico* profilers to be developed for category formation is of key importance. The approach requires extensive work to identify AOPs for a wide range of regulatory endpoints and then the development of additional *in vitro* assays that can be used to investigate the key mechanistic steps. Finally, the combination of existing *in vivo* data and new *in vitro* data can be used in conjunction with one another to develop *in silico* tools to enable chemicals to be grouped into mechanism-based categories. This approach has the advantage in that the chemicals grouped into these categories are both chemically and biologically similar. Clearly, much work is needed in order to develop the AOP concept for other important regulatory endpoints.

2.8 Conclusions

The key step in the development of a category is in the definition of an appropriate measure of similarity by which to group the chemicals. Three key methods exist, these being; simple organic functional group based analogues, mechanism-based similarity and chemoinformatic measures of similarity. All three methods have been shown, in the literature, to be useful. However, the most powerful of these methods is mechanism-based similarity in which an understanding of (at least) the molecular initiating event is used to develop *in silico* profilers able to group chemicals. Such methods require expert knowledge in order to define structural alerts related to molecular initiating events. This expert knowledge is derived from a combination of *in vivo*, *in vitro* and *in chemico* data sources. Currently, *in silico* profilers are well developed in which the formation of a covalent adduct with a biological macromolecule acts as the molecular initiating event. In contrast, such profilers are less well developed for molecular initiating events that are receptor mediated. Finally, the concept of AOPs as a future direction for chemical category formation has been outlined. This approach, whilst currently in its infancy, is a road map to the development of *in silico* methods for grouping chemicals in terms of their chemical and biological similarity.

Acknowledgements

Funding from the European Community's Seventh Framework Program (FP7/2007–2013) COSMOS Project under grant agreement n°266835 and from Cosmetics Europe is gratefully acknowledged. Funding from the European Chemicals Agency (Service Contract No. ECHA/2008/20/ECA.203) is also gratefully acknowledged.

References

1. M. D. Barratt, QSAR, read-across and REACH. *ATLA*, 2003, **31**, 463.
2. Guidance on grouping of chemicals (available from: http://www.oecd.org).
3. Guidance on information requirements and chemical safety assessment Chapter R.6: QSARs and grouping of chemicals (*available from:* http://guidance.echa.europa.eu/).
4. T. W. Schultz, (2010) Adverse outcome pathways: A way of linking chemical structure to *in vivo* toxicological hazards, in *In silico Toxicology: Principles and Applications*, ed. M. T. D. Cronin and J. C. Madden, Royal Society of Chemistry, Cambridge, UK. 2010, p. 346.
5. A. O. Aptula, G. Patlewicz and D. W. Roberts, Skin sensitisation: Reaction mechanistic applicability domains for structure-activity relationships, *Chem. Res. Toxicol.*, 2005, **18**, 1420.
6. D. W. Roberts, G. Patlewicz, P. S. Kern, F. Gerberick, I. Kimber, R. J. Dearman, C. A. Ryan, D. A. Basketter and A. O. Aptula, Mechanistic applicability domain classification of a local lymph node assay dataset for skin sensitisation, *Chem. Res. Toxicol.*, 2007, **20**, 1019.
7. S. J. Enoch, C. M. Ellison, T. W. Schultz and M. T. D. Cronin, A review of the electrophilic reaction chemistry involved in covalent protein binding relevant to toxicity, *Crit. Rev. Toxicol.*, 2011, **41**, 783.
8. S. J. Enoch, D. W. Roberts and M. T. D. Cronin, Electrophilic reaction chemistry of low molecular weight respiratory sensitisers, *Chem. Res. Toxicol.*, 2009, **22**, 1447.
9. S. J. Enoch, M. J. Seed, D. W. Roberts, M. T. D. Cronin, S. J. Stocks and R. M. Agius, Development of mechanism-based structural alerts for respiratory sensitisation hazard identification, *Chem. Res. Toxicol.*, 2012, **25**, 2490.
10. National-Research-Council, Scientific frontiers in developmental toxicology and risk assessment, National Academy Press, Washington, 2000.
11. E. Fabjan, E. Hulzebos, W. Mennes and A. H. Piersma, A Category Approach for Reproductive Effects of Phthalates, *Crit. Rev. Toxicol.*, 2006, **36**, 695.
12. S. J. Enoch, M. T. D. Cronin and C. M. Ellison, The use of a chemistry based profiler for covalent DNA binding in the development of chemical categories for read-across for genotoxicity, *ATLA*, 2011, **39**, 131.
13. S. J. Enoch, M. T. D. Cronin, T. W. Schultz and J. C. Madden, Quantitative and mechanistic read across for predicting the skin sensitisation potential of alkenes acting via Michael addition, *Chem. Res. Toxicol.*, 2008, **21**, 513.
14. D. W. Roberts, A. O. Aptula and G. Patlewicz, Mechanistic applicability domains for non-animal based prediction of toxicological endpoints. QSAR analysis of the Schiff base applicability domain for skin sensitisation, *Chem. Res. Toxicol.*, 2006, **19**, 1228.

15. S. J. Enoch, D. W. Roberts and M. T. D. Cronin, Mechanistic category formation for the prediction of respiratory sensitisation, *Chem. Res. Toxicol.*, 2010, **23**, 1547.
16. S. J. Enoch and M. T. D. Cronin, A review of the electrophilic reaction chemistry involved in covalent DNA binding, *Crit. Rev. Toxicol.*, 2010, **40**, 728.
17. P. K. Schmieder, G. T. Ankley, O. Mekenyan, J. D. Walker and S. Bradley, Quantitative structure-activity relationship models for prediction of estrogen receptor binding affinity of structurally diverse chemicals, *Environ. Toxicol. Chem.*, 2003, **22**, 1844.
18. G. Patlewicz, N. Jeliazkova, A. Gallegos Saliner and A. P. Worth, Toxmatch - a new software tool to aid in the development and evaluation of chemically similar groups, *SAR QSAR Environ. Res.*, 2008, **19**, 397.
19. J. Jaworska and N. Nikolova-Jeliazkova, How can structural similarity analysis help in category formation? *SAR QSAR Environ. Res.*, 2007, **18**, 195.
20. A. R. Leach, *Molecular modelling: principles and applications*, Pearson Education Limited Harlow, 2001, 640.
21. V. J. Gillett and A. R. Leach, *An Introduction to Chemoinformatics*. Kluwer Academic Publishers Dordrecht, The Netherlands, 2003, 103.
22. S. J. Enoch, M. T. D. Cronin, J. C. Madden and M. Hewitt, Formation of structural categories to allow for read-across for teratogenicity, *QSAR Combi. Sci.*, 2009, **28**, 696.
23. S. J. Enoch, J. C. Madden and M. T. D. Cronin, Identification of mechanisms of toxic action for skin sensitisation using a SMARTS pattern based approach, *SAR QSAR Environ. Res.*, 2008, **19**, 555.
24. A. O. Aptula, G. Patlewicz, D. W. Roberts and T. W. Schultz, Non-enzymatic glutathione reactivity and in vitro toxicity: A non-animal approach to skin sensitisation, *Toxicol. In Vitro*, 2006, **20**, 239.
25. D. W. Roberts, A. O. Aptula and G. Patlewicz, Electrophilic chemistry related to skin sensitisation. Reaction mechanistic applicability domain classification for a published data set of 106 chemicals tested in the mouse local lymph node assay, *Chem. Res. Toxicol.*, 2007, **20**, 44.
26. S. J. Enoch, M. T. D. Cronin and T. W. Schultz, The definition of the applicability domain relevant to skin sensitisation for the aromatic nucleophilic substitution electrophilic mechanism, *SAR QSAR Environ. Res.*, 2012, **23**, 649.
27. H. J. M. Verhaar, C. J. van Leeuwen and J. L. M. Hermens, Classifying environmental pollutants. 1: Structure-activity relationships for prediction of aquatic toxicity, *Chemosphere*, 1992, **25**, 471.
28. P. C. von der Ohe, R. Kuhne, R. U. Ebert, R. Altenburger, M. Liess and G. Schüürmann, Structural alerts - A new classification model to discriminate excess toxicity from narcotic effect levels of organic compounds in the acute daphnid assay, *Chem. Res. Toxicol.*, 2005, **18**, 536.

29. M. Nendza and M. Muller, Discriminating toxicant classes by mode of action: 2. Physico-chemical descriptors, *Quant. Struct.-Act. Relat.*, 2001, **19**, 581.
30. M. Nendza and M. Muller, Discriminating toxicant classes by mode of action: 3. Substructure indicators, *SAR QSAR Environ. Res.*, 2007, **18**, 155.
31. M. Nendza and A. Wenzel, Discriminating toxicant classes by mode of action - 1. (Eco)toxicity profiles. *Environ. Sci. Pollut. Res.*, 2006, **13**, 192.
32. D. W. Roberts, Linear free-energy relationships for reactions of electrophilic halobenzenes and pseudohalobenzenes, and their application in prediction of skin sensitisation potential for S_NAr electrophiles, *Chem. Res. Toxicol.*, 1995, **8**, 545.
33. D. W. Roberts and D. A. Basketter, Quantitative structure-activity relationships: sulfonate esters in the local lymph node assay, *Contact Dermatitis*, 2000, **42**, 154.
34. D. W. Roberts, B. F. J. Goodwin, D. L. Williams, K. Jones, A. W. Johnson and J. C. E. Alderson, correlations between skin sensitisation potential and chemical-reactivity for para-nitrobenzyl compounds, *Fd. Chem. Toxicol.*, 1983, **21**, 811.
35. D. W. Roberts and A. Natsch, High throughput kinetic profiling approach for covalent binding to peptides: Application to skin sensitisation potency of Michael acceptor electrophiles, *Chem. Res. Toxicol.*, 2009, **22**, 592.
36. A. O. Aptula, D. W. Roberts and M. T. D. Cronin, From experiment to theory: Molecular orbital parameters to interpret the skin sensitisation potential of 5-chloro-2-methylisothiazol-3-one and 2-methylisothiazol-3-one, *Chem. Res. Toxicol.*, 2005, **18**, 324.
37. A. O. Aptula, D. W. Roberts, T. W. Schultz and C. Pease, Reactivity assays for non-animal based prediction of skin sensitisation potential, *Toxicology*, 2007, **231**, 117.
38. T. W. Schultz, K. Rogers and A. O. Aptula, Read-across to rank skin sensitisation potential: subcategories for the Michael acceptor domain, *Contact Dermatitis*, 2009, **60**, 21.
39. D. W. Roberts, T. W. Schultz, E. M. Wolf and A. O. Aptula, Experimental reactivity parameters for toxicity modelling: Application to the acute aquatic toxicity of S_N2 electrophiles to *Tetrahymena pyriformis*, *Chem. Res. Toxicol.*, 2010, **23**, 228.
40. T. W. Schultz, T. I. Netzeva, D. W. Roberts and M. T. D. Cronin, Structure-toxicity relationships for the effects to *Tetrahymena pyriformis* of aliphatic, carbonyl-containing, α, β-unsaturated chemicals, *Chem. Res. Toxicol.*, 2005, **18**, 330.
41. T. W. Schultz, K. E. Ralston, D. W. Roberts, G. D. Veith and A. O. Aptula, Structure-activity relationships for abiotic thiol reactivity and aquatic toxicity of halo-substituted carbonyl compounds, *SAR QSAR Environ. Res.*, 2007, **18**, 21.

42. S. J. Enoch and M. T. D. Cronin, Development of new structural alerts for chemical category formation for assigning covalent and non-covalent mechanisms relevant to DNA binding, *Mutat. Res. – Gen. Toxicol. Environ. Mutagen.*, 2012, **743**, 10.
43. T. W. Schultz, R. E. Carlson, M. T. D. Cronin, J. L. M. Hermens, R. Johnson, P. J. O'Brien, D. W. Roberts, A. Siraki, K. B. Wallace and G. D. Veith, A conceptual framework for predicting the toxicity of reactive chemicals: modeling soft electrophilicity, *SAR QSAR Environ. Res.*, 2006, **17**, 413.
44. K. Landsteiner and J. Jacobs, Studies on the sensitisation of animals with simple chemical compounds, *J. Exp. Med.*, 1935, **61**, 643.
45. K. Landsteiner and J. Jacobs, Studies on the sensitisation of animals with simple chemical compounds II, *J. Exp. Med.*, 1936, **64**, 625.
46. T. W. Schultz, J. W. Yarbrough, R. S. Hunter and A. O. Aptula, Verification of the structural alerts for Michael acceptors, *Chem. Res. Toxicol.*, 2007, **20**, 1359.
47. F. Bajot, M. T. D. Cronin, D. W. Roberts and T. W. Schultz, Reactivity and aquatic toxicity of aromatic compounds transformable to quinone-type Michael acceptors, *SAR QSAR Environ. Res.*, 2011, **22**, 51.
48. J. A. H. Schwöbel, Y. K. Koleva, S. J. Enoch, F. Bajot, M. Hewitt, J. C. Madden, D. W. Roberts, T. W. Schultz and M. T. D. Cronin, Measurement and estimation of electrophilic reactivity for predictive toxicology, *Chem. Rev.*, 2011, **111**, 2562.
49. T. W. Schultz and J. W. Yarbrough, Trends in structure-toxicity relationships for carbonyl-containing α,β-unsaturated compounds, *SAR QSAR Environ. Res.*, 2004, **15**, 139.
50. S. J. Enoch, M. T. D. Cronin and T. W. Schultz, The definition of the toxicologically relevant applicability domain for the $S_N Ar$ reaction for substituted pyridines and pyrimidines, *SAR QSAR Environ. Res.*, **24** (5), 651.
51. A. Natsch and H. Gfeller, LC-MS-based characterisation of the peptide reactivity of chemicals to improve the *in vitro* prediction of the skin sensitisation potential, *Toxicol. Sci.*, 2008, **106**, 464.
52. G. T. Ankley, R. S. Bennett, R. J. Erickson, D. J. Hoff, M. W. Hornung, R. D. Johnson, D. R. Mount, J. W. Nichols, C. L. Russom, P. K. Schmieder, J. A. Serrano, J. E. Tietge and D. L. Villeneuve, Adverse outcome pathways: A conceptual framework to support ecotoxicology research and risk assessment, *Environ. Toxicol. Chem.*, 2010, **29**, 730.

CHAPTER 3
Informing Chemical Categories through the Development of Adverse Outcome Pathways

K. R. PRZYBYLAK*[a] AND T. W. SCHULTZ[b]

[a] Liverpool John Moores University, School of Pharmacy and Chemistry, Byrom Street, Liverpool L3 3AF, England; [b] University of Tennessee, College of Veterinary Medicine, 2407 River Drive, Knoxville, TN 37996, USA
*E-mail: k.r.przybylak@ljmu.ac.uk

3.1 Introduction

In general, chemical categories are formed based on chemical similarity or the potential for a particular chemical-biological interaction at the molecular level – referred to as a molecular initiating event (MIE) (see Chapter 2). For regulatory purposes, it would be most useful to build a chemical category based on the *in vivo* endpoint considered in the assessment. However, this can be achieved only in cases where the *in vivo* endpoint is related directly (sometimes called "hard wired") to the MIE. Therefore, it is essential to develop a causal linkage between the interaction of a chemical with a biomolecule at the molecular level and the subsequent biological effects at the subcellular, cellular, tissue, organ, whole animal and population levels. To this end, the concept of the Adverse Outcome Pathway (AOP) has been introduced recently to provide such a mechanistically plausible and transparent link between MIEs and the *in vivo* outcomes of regulatory interest.[1,2] Moreover, the AOP concept provides a useful structure within which existing knowledge from *in vivo* tests can be integrated with the results of other methods including: molecular screening and omics assays; computationally-based predictions; as well as advances in bioinformatics and systems biology. Such collated information related to the pathway is then organised at different levels of biological organisation and/or other seminal dimensions (*e.g.* species, gender, life stage). In that form, the AOP is intended to inform chemical category formation, particularly with the aim of filling data gaps by read-across.

The AOP concept is the continuation of the generally accepted and much advocated pathway-based approaches, such as dose–response models,[3,4] Mode-of-Action (MoA) framework[5] and toxicity pathway concept.[6] The pathway approach is based on the idea that toxicity results from exposure to the chemical and a molecular interaction with an initial key target such as a protein or receptor in the organism. The goal of the pathway concept is to improve transparency and efficiency, and decrease uncertainty in the decision making process. As a pathway-based approach, an AOP is the sequential progression of events from the MIE to the *in vivo* outcome of interest (Figure 3.1). Generally, it refers to a broader set of pathways that would:

- proceed from the MIE, in which a chemical interacts with a biological target (*e.g.* DNA binding, protein oxidation, or receptor/ligand interaction *etc.*);
- continue on through a cascade of biological activities (*e.g.* gene activation, altered cellular chemistry or tissue development *etc.*);
- ultimately culminate in an adverse effect of relevance to human health or ecological risk assessment (*e.g.* mortality, disrupted reproduction, cancer, extinction *etc.*).[7]

While AOPs may be depicted as linear frameworks (see Figure 3.1), toxicity is multi-dimensional, therefore the pathway between a MIE and the apical adverse effect can vary significantly. This is especially true for human health endpoints, where effects are the result of multiple organ interactions (*e.g.*, skin sensitisation), multiple events (*e.g.*, repeat dose toxicity), accumulation over time (*e.g.*, neural toxicity), or are related to a specific life stage of an organism (*e.g.*, developmental toxicity).

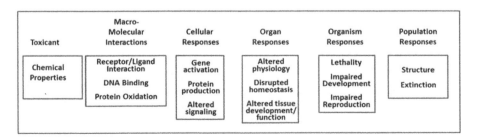

Figure 3.1 A schematic representation of an AOP illustrated with reference to a number of pathways (adapted from Ankley *et al.*[2]).

AOPs are typically represented as a sequence of key events, moving from one key event to another. In such a form, the AOP can be applied as a "bottom-up" approach, whereby chemical and mechanistic information is used to define MIEs that can inform chemical category formation and can then be applied in hazard assessment. Conversely, an AOP can also be used in a "top-down" approach by taking the apical adverse outcomes produced by well-studied compounds and establishing a MoA. This information can then be used to develop chemical categories. The chemical categories developed based on an AOP allow for the transition from a chemical-by-chemical approach to a chemical-category approach and therefore reduce financial cost, time and use of test animals. On the other hand, AOPs provide a consistent structure for organising toxicological knowledge across levels of biological organisation and can aid in identifying gaps in that knowledge and thus drive the development of alternative testing methods to fill such gaps.[2]

The aim of this chapter, therefore, is to provide the basis for the rapidly expanding field of AOP led analysis of toxicological data. This is achieved by providing explanations and definitions of AOPs and specifically how they inform the development of categories to allow for read-across. The information that may be obtained from an AOP in terms of category is illustrated by a case study.

3.2 The Structure of the AOP

The AOP concept describes events that occur following chemical exposure. AOPs have to be developed bearing in mind that the chemical will induce toxicity only when the exposure (of sufficiently high level and/or duration) will exceed the adaptive response of an organism. The pathway begins with the interaction of the chemical with a biomolecule, which is followed by sequential perturbations at the cellular, tissue and organ-level that lead to the adverse effect of interest. As such, the AOP consists of three main information blocks: the MIE, intermediate events and the final or apical adverse effect. In any case, a given apical endpoint will be associated with a finite set of possible MIEs. Similarly, a given MIE will be associated with a finite set of possible apical endpoints. However, each AOP will have only one MIE and one apical endpoint, which could be related to an adverse effect at higher levels of organisation.

3.2.1 Development of the AOP

The development of the AOP can be started from any of the three blocks, depending on what knowledge is available at the beginning of the exercise. Typically AOP development begins with either the MIE or the final adverse effect of interest. The latter reflects the fact that an AOP is anchored at its two ends by the chemical/biological interaction and outcome of interest (Figure 3.2).

Informing Chemical Categories through Adverse Outcome Pathways 47

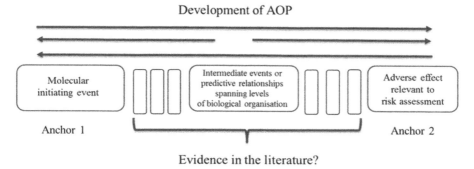

Figure 3.2 A schematic diagram for the development of an AOP starting at any of the three main blocks of information (adapted from OECD).[8]

The MIE explains the nature of the chemical interaction with biological (macro)molecules. This information allows for an initial description of the molecular structure limitations for chemical category members acting in a similar manner. The identification of the adverse effect relevant to the assessment is another crucial aspect in the development of the AOP. It is essential to define this adverse effect clearly, as it determines the most relevant mechanistic information which helps to define the third block of the AOP — the intermediate effects.

To develop the AOP, different types of data can be utilised, these include: structural alerts that reflect the types of chemicals that can initiate a pathway (discussed in Chapters 2 and 6); *in chemico* methods that measure the relative reactivity or chemical-biological interactions; *in vitro* assays that confirm the subsequent cellular responses (*e.g.* gene expression); and, ultimately, *in vivo* tests that measure endpoints that are directly relevant to the adverse effect that drives regulatory decision making.[9] It has to be stressed that before data can be used in the development of an AOP, there should be a framework for accepting these data based on a set of quality criteria (see Chapter 5). Moreover, there must be reassurance that the data upon which the AOP is based are derived from carefully designed experiments using appropriate dose/exposure levels. The data gathered can be used to identify key steps in the AOP and provide scientific evidence to support the AOP.

The development of AOPs also requires multidisciplinary collaboration involving experts in toxicology, chemistry and biology, all of whom may use different terminology. This can lead to confusion among scientists and organisations. Therefore a standardised set of terminology has been developed to assist in the understanding of the AOP concept as well as its assessment, recording and ultimate acceptance.[7] Moreover, the use of a common ontology also helps the application of the AOP concept in developing quantitative structure–activity relationship (QSAR) models and chemical categories to advance the use of predictive techniques in risk assessments. The Organisation for Economic Co-operation and Development (OECD) compiled terms related to the AOP concept.[8] As a result, 42 terms, the majority of which had multiple definitions, have been collected from the literature.

3.2.1.1 Identification of the Adverse Effect

The identification of an adverse effect after xenobiotic exposure has been a mainstay for assessing risk to inform risk management decisions.[10] An adverse effect can be defined based on different levels of biological organisation: cellular/tissue, organ, organ system, individual, population or ecosystem. Usually, the apical endpoint is associated with an *in vivo* response as indicated in standard test guidelines. However, in certain cases, such as cell proliferation or bioenergetics, the apical endpoint may be at a lower level of biological organisation. Moreover, the adverse effects can cover long term health endpoints as well as local effects. In the former case, the adverse effects are the results of multiple events (*e.g.* repeat dose toxicity) or accumulation over time (*e.g.* neural toxicity) or are related specifically to a particular life stage of the organism (*e.g.* developmental toxicity). In the second scenario, MIEs are likely to be closely aligned with the *in vivo* outcome (*e.g.* skin sensitisation, skin and eye irritation). It is essential to clearly and precisely define the adverse effect as one of the anchors of the AOP. This helps to define the mechanistic sequence of events leading to this outcome.

3.2.1.2 Definition of the Molecular Initiating Event (MIE)

The MIE described as the chemical-induced perturbation of biological systems at the molecular level represents the primary anchor of the AOP. Therefore, it is very important to identify clearly the beginning of the cascade leading to the specified adverse effect. Many MIEs have been defined in terms of covalent binding to proteins and/or DNA (see Chapters 2 and 6).[11] These types of MIEs are based on the principles of organic chemistry (*i.e.* electrophile–nucleophile reactivity). In contrast, MIEs may also be based on interactions which are "receptor binding", or binding to enzymes, these are often non-covalent and more selective in nature.[12] Understanding the MIE allows for the definition of the properties of chemicals inducing the perturbation, such as bioavailability, structural requirements (especially for receptor binding) and metabolic transformation. For example, potential inducers of respiratory impairment caused by the non-covalent perturbation of the inner mitochondrial membrane are characterised as phenolic weak acid uncouplers, such as polyhalogen-substituted (three or more), dinitro-substituted or polyhalogen- and mono-nitro-substituted phenols with log P between 1.5 and 5.5 and a pKa value between 3 and 6.[7] Therefore, the understanding of the chemistry of potential inducers helps to define the molecular structure limitations for chemical category members acting via a similar mechanism.

3.2.1.3 Recognition of Key Events Leading to the Adverse Effect

The AOP must describe a pathway that has key events that are known to be feasible in normal physiological and/or biochemical processes. Therefore, it is crucial to understand the basis of normal physiology (*e.g.* reproductive processes, nervous system function, liver functions *etc.*) before being able to identify intermediate events that lead to an adverse effect. This will help in recognising the complex networks of processes, at different levels of biological organisation, which can be disrupted. The identified AOP must not contradict any steps of normal biological processes. During the identification of key steps, a review of the existing literature is required to find out as much information as possible about a plausible mechanism and the intermediate steps leading to the final adverse effect. This aspect is crucial for the development of the AOP. It requires manual evaluation of the scientific literature to determine relevant intermediate events and their usefulness as key events in developing the AOP. Usually multiple intermediate events are identified. Therefore, the assembled knowledge has to be filtered and selected to match the single AOP.

To be a key event, the intermediate step must be capable of being evaluated experimentally. That is to say, the event must be able to be used in a hypothesis which can then be tested. For instance, dendritic cell maturation observed during skin sensitisation can be characterised by the expression of specific cell surface markers, such as adhesion molecules, chemokines and cytokines, which can be measured by assays such as the *in vitro* VITROSENS test.[13] This assay, using human CD34+ progenitor-derived dendritic cells (CD34-DC), is based on the differential expression of the cAMP-responsive element modulator (CREM) and monocyte chemotactic protein-1receptor (CCR2) which are able to discriminate between skin sensitisers and non-sensitisers. This fact is of key importance for category formation, as it allows for the testing of potential category members to ensure that they follow the pathway described by the AOP.

3.2.2 The Assessment of the AOP

During the development of an AOP, it is considered critical to be able to gauge its reliability and robustness. This should be done by evaluating the experimental support of the AOP. In such assessment, the qualitative and quantitative understanding of the AOP has to be analysed. This means that every key step should be clearly identified and documented with relevant scientific evidence and its evaluation. For the quantitative understanding of an AOP, the threshold and scale of the linkage between key events in the pathway play important roles. Moreover, the assessment of the quantitative understanding of an AOP should determine the response-to-response relationships required to scale *in vitro* effect(s) to *in vivo* outcomes. As the AOP is supported by various data, it is vital to assess the Weight-of-Evidence (WoE) supporting the AOP. It can be done by implementation of the Bradford Hill criteria, which help to evaluate the relevance of the scientific evidence gathered to the hypothesised AOP.[14] The aspects considered include concordance, strength and consistency between the adverse effect, initiating event and key events, biological plausibility, coherence and consistency of the experimental evidence, as well as uncertainties, inconsistencies and data gaps.

The final step in the assessment of an AOP is a statement regarding the confidence associated with it. Confidence in an AOP is increased by a more comprehensive understanding of the nature of the interaction between the chemical and the biological system, coupled with mechanistic understanding of the biological response.

3.3 Harmonised Reporting and Recording of an AOP

Where possible the information collected should be used to present the whole adverse pathway step-by-step. This means starting from the characterisation of the route of exposure and chemical properties and the identification of the molecular initiating event and site of action. After that, the responses at the molecular, cellular/tissue, organ, organism and population/ecosystem levels should be identified. The final stage depends on the level of biological organisation of the adverse outcome. This information should be reported systematically and in a transparent way to maintain consistency among the developed AOPs. The existing AOPs' documents show the lack of the standardisation of procedures during the development and documentation process.[7,9,15–19] Analysis of these documents showed significant differences in the documentation of the AOPs. Different levels of information are available among these reports, however, for some of them, no clear assessment of the AOP is made. Therefore, it is important to provide the framework for consistent information gathering and organisation into an AOP. The template should give an insight into which pieces of information are necessary to identify an AOP and how to present them. It will also provide initial assistance on how to undertake the assessment of an AOP in terms of its completeness and relevance. Such a template has been recently proposed by the OECD;[8] this was developed based on the AOP already established for skin sensitisation initiated by covalent binding to proteins.[18,19] According to this guidance, every key step has to be identified and supported by experimental evidence together with an evaluation of the evidence. The summary should be presented in the form of a table and a graphical diagram and finally a clear assessment of the AOP has to be undertaken.

Once the AOPs are developed and assessed, they should be stored in a publicly available repository to provide easy access to these documents. It is important as AOPs are living documents that will be continually updated and refined as more data are generated, analysed, and developed within the AOP process. Wiki-based online environments such as Effectopedia[20] or the World Health Organization (WHO) MoA Tool[21] can be used, not only as repositories, but also they can assist in the development of AOPs to improve the effectiveness of communication among experts from diverse areas of science who may not naturally communicate with each other. This improved interaction is critical to the kind of synthesis of knowledge about chemical interactions, metabolism, systems biology and ecology needed for AOP development.

3.4 Use and Benefits of an AOP

A well-identified AOP, with an accurately described sequence of events through the different levels of biological organisation, provides valuable pieces of mechanistic information which can be used for many purposes.[22] The major advantage of the AOP is the provision of the transparent causal linkage between the MIE and the final *in vivo* outcome of interest. Moreover, the AOP is designed to avoid mixing information from multiple mechanisms (*i.e.* different molecular initiating events which can cause the same *in vivo* outcome through different AOPs). Table 3.1 presents a list of existing AOPs together with the associated chemically induced perturbations at the molecular level. This is crucial information for the developing mechanistic profilers within the OECD (Q)SAR Toolbox to improve the grouping methods.[23] It helps to form chemical categories based on toxicological behaviour that are especially useful for the prediction of long term and chronic effects.

3.4.1 Developing Chemical Categories Supported by an AOP

The most likely and originally intended application of AOPs, even incomplete ones, is to inform chemical grouping strategies or chemical category formation and the development of structure-activity relationships (SARs). An important

Table 3.1 The AOP's linkage between the *in vivo* outcome and MIE.

Endpoint	MIE/Profiler
Skin sensitisation[18,19]	Protein binding
Aquatic toxicity/Non-polar narcosis[2]	Hydrophobic interaction with neuronal membrane
Aquatic toxicity/Photoactivated toxicity[2]	Reactive single oxygen formation
Aquatic toxicity/Aryl hydrocarbon receptor[2]	Aryl hydrocarbon receptor binding
Reproductive toxicity[2,7]	Oestrogen receptor binding
Reproductive toxicity[2]	Aromatase inhibition
Neurotoxicity[16]	Kainate receptor binding
Repeat dose/Haemolytic anaemia[24]	Metabolic activation of nitrobenzenes releasing Reactive Oxygen Species
Nephrotoxicity[7]	Metabolic activation of 4-aminophenols binding to glutathione
Respiratory impairment/Weak Acid Respiratory Uncouplers[7]	Non-covalent perturbation by weak acids leading to loss of H^+ gradient

advantage of the AOP approach in chemical grouping is that it allows categorisation of chemicals based on toxicological similarity. Considering not only the initial interaction of a chemical at the molecular level, but also the perturbations at higher biological levels, provides an opportunity to group chemicals based on both intrinsic chemical and biological activity. Such categorisation of chemicals based on both MIE and early key events gives greater confidence that chemicals induce adverse effect via the same toxicity mechanism.

For more effective use of AOPs in developing chemical categories, three libraries of information should be collated and integrated (the examples given in parentheses relate to skin sensitisation):

- a library of *in vivo* effects typically used in assessments (*e.g.* EC3 values in the local lymph node assay);
- a library of molecular initiating events (*e.g.* protein binding reactivity);
- a library of intermediate events, typically data generated using *in vitro* methods (*e.g.* dendritic cell surface biomarkers).

Each library can, in theory, be associated with a single, or multiple, chemical domain(s). With regard to chemical categories, the chemical structure space covered, or applicability domain, is reliant on the chemicals assessed for the MIEs and the key events within the AOP.

Once an AOP is identified and understood for one compound, it becomes possible to identify other chemicals that perturb elements of the AOP, using relatively inexpensive and rapidly performed techniques, such as high throughput screening (HTS), SAR, and microarrays. AOPs are likely to be useful to predict the potential for less well studied chemicals to induce an adverse outcome of interest. For example, if a chemical is shown to elicit a particular MIE associated with a given AOP and subsequent key events, the chemical may be predicted with some confidence to produce the same adverse outcome, even if the definitive *in vivo* test has not been conducted. Moreover, such application of an AOP in the formation of a chemical category and filling of data gaps by read-across could drive more rational testing. Thousands of chemicals could be assessed for their ability to elicit the MIE and perhaps one, or a few, key events at the cellular level. Subsequently, hundreds of chemicals could be assessed for key events relevant to higher levels of biological organisation and in this manner only a few chemicals (maybe only tens) would need to be assessed for the final *in vivo* adverse effect. In this way chemical category formation and read-across could be informed by the thousands of chemicals pre-screened for the MIE(s). Subsequently, the chemical category could be refined and sub-categorised based on the information provided by the chemicals screened for selected key events. Ultimately, the data gap could be filled by read-across from the relevant chemicals assessed for the final *in vivo* endpoint.

3.4.2 General Applications of AOP for Regulatory Purposes

From the regulatory point of view, an AOP can be used in a variety of applications.[22] These will depend on the type of decision to be made and what level of uncertainty is acceptable for that assessment. These decisions can include:

- priority setting for further testing;
- hazard identification;
- classification and labelling;
- risk assessment.

For the first two applications, a partial AOP, where not all key events are known, can be used. In the case of complete risk assessment, a quantitative AOP with dose–response information as well as absorption, distribution, metabolism, and excretion (ADME) properties will be required.

A well-identified AOP, with an accurately described sequence of events through the different levels of biological organisation, provides valuable pieces of mechanistic information which can be used to inform the work bodies such as the OECD Test Guideline Programme. Based on well identified and scientifically proven key events, new *in vitro, ex vivo* and HTS assays that detect direct chemical effects or responses at the cellular or higher levels of biological organisation can be developed.[7] In addition, an AOP, for any given hazard endpoint, can be the basis for developing an integrated approach to testing and assessment or an integrated testing strategy for that hazard endpoint.

3.5 A Case Study: Developing a Chemical Category for Short-Chained Carboxylic Acids Linked to Developmental Toxicity

The following section outlines how to develop a chemical category for short-chained carboxylic acids (SCCAs) through the development of the AOP for developmental toxicity. To map out the AOP, firstly an overview of the process of developmental toxicity is required. This is the first part of this section which is given below. In order to illustrate how this can support the development of categories, and hence read-across techniques, results from the testing of a group of compounds known to cause developmental toxicity are presented.

3.5.1 Overview of Developmental Toxicity

The development of embryonic structures results from a well-orchestrated series of complicated molecular, biochemical, and physical events that eventually lead to what we recognise as a fully developed infant.[25] These events are highly conserved across vertebrates, especially mammals. While the development of a structure begins with cell proliferation other events (*e.g.* cell differentiation and migration) are also critical. Early embryonic primordia are acted upon by inductive forces. Such forces or signals promote cellular differentiation, thereby providing the necessary anlagen (the clustering of embryonic cells) for the development of the final embryonic structures. Many of the inductive signals are secreted growth factors passed between cell populations of interacting tissues.[25]

Ontogenetic development in humans and other vertebrates, whilst a continuum, is often divided into a series of stages. Briefly, the pre-differentiation period is associated with development from the single cell zygote to the tri-layered (*i.e.*, ectoderm, endoderm, and mesoderm) gastrula. The early differentiation period of development is characterised by organogenesis (*i.e.*, the development of the organ systems), while the advanced differentiation period is the time of morphogenesis and histogenesis (*i.e.*, the development of body form and tissues). Fœtalgenesis and, in some species including man, the early postnatal period is the time of initial growth and the acquisition of full organ function, whilst the adult period can be considered the remainder of life.

Manifestations in developmental toxicity include the types, degree, and phenotypic incidences of abnormal development. Final manifestations of developmental toxicity include: death, malformation, growth retardation, and functional deficit. Life stage and manifestations of developmental toxicity are linked. An early refractory period is associated with pre-differentiation; the totipotency of cells (the so-called stem cells) present during cleavage and blastulation and the pluropotency of the early germ layer maturation mean one cell can readily replace another. The terminal juncture of gastrulation is associated with the formation of third germ layer (*i.e.* mesoderm) and is typically the most sensitive time in life. At this point in development, cells, which were multi-potent become less so. For example, the development of the nervous system and other ectodermal elements become separated from (but influenced by) development of muscle, connective tissue and other mesodermal elements. Subsequently, during histogenesis and organogenesis, and cellular specialisation, potency decreases further so hepatocytes (liver) are not capable of becoming pneumocytes (lung).

High susceptibility to malformation is associated with the period of histogenesis and organogenesis. Interference with development during organogenesis often leads to organ and system specific phenotypic effects. For example, the nervous system, which is the first system to develop, is susceptible early in organogenesis, while the urinary and reproductive systems, which are the last systems to develop, are susceptible late in organogenesis. The period of growth and function susceptibility is associated with the fœtal period and early postnatal development. Interference with development at this period typically results in growth retardation and organ-specific functional disturbances including neural toxicity.

Major developmental effects also are life stage-dependent. For example, death is the most common outcome associated with exposure during early embryogenesis (up through gastrulation), malformation is associated with exposure during organogenesis, and growth retardation and functional deficiency is associated with exposure during fœtalgenesis.

One of the most critical of the signals between two opposing cell populations involves epithelial–mesenchymal interactions. The mesenchymal cells influence the epithelium, which can differentiate and secrete factors that subsequently influence the mesenchyme. Such interactions continue until a target organ develops with organ-specific cell populations.

3.5.2 Valproic and Other Short-Chained Carboxylic Acids as Developmental Toxicants

It has long been know that a disproportionate number of known developmental toxicants are weak acids. Among these weak acids is the well-studied SCCA: 2-propylpentanoic acid known as valproic acid.[25] In humans, *in utero* exposure to valproic acid has been linked to neural, craniofacial, cardiovascular and skeletal defects with the developing nervous system appearing to be particularly sensitive.

Dose–response data are available for the effects of valproic acid from a variety of *in vivo*, *ex vivo* and *in vitro* investigations. Except for reduced sensitivity of rodents *in vivo*, all species show developmental and neurogenic effects. As an example, valproic acid treatment in mice on day 8–9 of gestation causes failure of the closure of the cranial end of the neural tube, spina bifida, and limb abnormalities.[26] In all embryo-based studies, once initiated, the adverse effects of valproic acid exposure on early development cannot be reversed.

Valproic acid activates the Wnt/β-catenin signalling pathway resulting in anti-proliferative and pro-differentiation effects on cell membranes, which impairs embryonic development. The Wnt pathway helps regulate β-catenin and cadherins in cells.[27] Up-regulation of β–catenin and down-regulation of E-cadherin in favour of N-cadherin during remodelling of cellular junctions leads to abnormally high levels of N-cadherin and neural tube cells fail to migrate. Specifically, valproic acid inhibits class I histone deacetyase.[28] Histone acetylation has been shown to be central to the regulation of gene expression in eukaryotes.[29] Histone acetylation is an epigenetic modification, which is balanced by the action of histone acetyl transferase and histone deacetylase. Histone deacetyases affect gene transcription and chromatin assembly by altering histones at the post-transcriptional level.

The cell responses to histone deacetylase inhibitors are complex, but generally an increase in the level of histone acetylation is associated with an increase in gene expression, which act as cell proliferation agents involved in the convergent extension (narrowing of a tissue in one axis and elongation in another) during development. As a consequence, inhibition of histone deacetylase impairs cell proliferation and differentiation. In very early development, the central nervous system is the chief target of deacetylase inhibitor-induced malformation.

A review of the literature, especially the more inclusive studies of Phiel et al.[30] and the review of Wiltse,[31] reveals that disruption of Wnt/β-catenin signalling is provoked by the inhibition of histone deacetylase. As the seminal event this inhibition can be regarded as the initial event of the mode of action leading to the adverse outcome of developmental toxicity or teratogenicity, especially for select SCCAs.

Based largely on evidence for valproic acid, a mode of action for chemicals acting as histone deacetylase inhibitors leading to developmental toxicity in humans is likely to take the following form:

1) The molecular site of action is the enzyme histone deacetylase.
2) The molecular initiating event is the inhibition of histone deacetylase, which leads to deacetylation of core histones.
3) The biochemical pathway is the alteration of Wnt-dependent transcription.
4) The first cellular level consequence is the accumulation of β-catenin, which leads to a reduction in E-cadherin in favour of N-cadherin.
5) The second cellular-level consequence is increased cell adhesion and concomitant reduced cell motility.
6) The target organ(s) or tissue(s) consequences depend on the development stage of exposure.
7) The physiological/anatomical response(s) to the cellular effects is prevention of convergent extension.
8) The embryonic response(s) to the biochemical, cellular, and physiological/anatomical effects is exhibited as specific terta (*e.g.* failure of the neural tube to elongate and close).
9) The overall effect on humans is embryo death and/or developmental impairment.

The scientific literature contains much information in support of this toxicity mechanism. Histone deacetylases control gene expression via the regulation of transcription. Valproic acid has been demonstrated to be a specific inhibitor of class I histone deacetylases.[28] In HeLa cells engineered to over-express the enzyme histone deacetylase, exposure to valproic acid resulted in the release of acetyl groups from acetylated histones in a dose–response relationship.[30] In human embryonic kidney cells (strain 293T) transfected with a luciferase bio-reporter system, exposure to valproic acid activated transcription in a dose–response relationship.[30] In mouse Neuro2A cells, exposure to valproic acid resulted in β-catenin being accumulated. When the Neuro2A cells were exposed to the protein synthesis inhibitor cycloheximide, β-catenin degraded. These results indicate that valproic acid promotes β-catenin synthesis but does not inhibit GSK-3β mediated degradation.

In vertebrate embryos, the neural crest is a provisional structure located at the dorso-lateral surface of the neural tube and consists of strips of ectodermal-derived cells. From these crests, cells become discontinuous, migrate to other locations, and form, amongst other things, the primordial for the dorsal or sensory root ganglia, somatic spinal nerves, and sympathetic ganglia of the autonomic nervous system with their visceral sympathetic nerves. In culture, neural tube segments of chick embryos exposed to valproic acid exhibit effects on the neural crest cells, which indicate interference with cell differentiation.[32] During normal development of neural crest cells, adhesion molecules and adherens junctions are lost or altered, the cytoskeleton is reorganised with the formation of cytoplasmic extensions, and the cells become motile. Upon exposure to valproic acid, there is a reduction in the number of neural crest cells, which become motile rather than cellular sheets being formed. In these sheets, N-cadherin is found in the cell boundaries. In contrast, in cells which remain independent and motile, N-cadherin is not found. Correspondingly, in mice with abnormally high levels of N-cadherin, cells fail to migrate from parts of the neural tube and spina bifida is observed.[32]

Valproic acid also disrupts early development of anterior structures of frog embryos.[33] Specifically, 88% of *Xenopus laevis* embryos exposed to 5 mM valproic acid for 24 hours at the mid-blastula stage of development exhibited a marked reduction in anterior structures and shortening of the anterior–posterior axis in embryos.[30] In *Xenopus*, dishevelled signalling via a planar cell polarity cascade is necessary for convergent extension of the neural tube.[34] The axis-inducing activity, stability, and sub-cellular distribution of β-catenin in *Xenopus* are regulated by the GSK-3β enzyme.[35] The Wnt/β-catenin pathway regulation of development of the central nervous system[36] explains the microcephaly in *Xenopus*.[33]

In cultured mouse embryos, valproic acid causes anterior neural tube defects, shortening of the anterior–posterior body axis, growth retardation, and failure of the embryo to rotate properly. In cultured rat embryos, it causes a decrease in embryonic growth, induction of neural tube defects and irregular somite formations and malformation in the fore- and mid-brain. Mechanistic understanding of these *in vitro* findings is provided by Wang *et al.*[37] who noted that disruption of dishevelled signalling results in defects related to central nervous system and stunting of the anterior–posterior axis.

Valproic acid is a known human teratogen.[38,39] Early *in vivo* studies have been reviewed by Di Carlos.[40] Later oral and subcutaneous dosing studies of pregnant mice and rats at 75 to 500 mg/kg valproic acid resulted in fœtuses exhibiting anterior neural tube defects, somite defects, heart malformations and spina bifida.[41]

3.5.3 AOP for Short-Chained Carboxylic Acids as Developmental Toxicants to Organisms in Aquatic Environments

From the mode of action and scientific literature noted above a number of key events considered essential to histone deacetylase inhibition induced convergent extension-related malformations can be identified and measured. These events and associated methods can form the basis of an AOP. Potential key intermediate events include:

1) The release of acetyl groups as a measure of inhibition of histone deacetylase.[42]
2) The accumulation of β-catenin as a measure of alteration of Wnt-dependent.[30,43]
3) Cell adhesion and/or cell motility as a measure of cell maturation and migration.[32]
4) Neural tube elongation and closing as a measure of convergent extension in early embryos.[30,33]

3.5.4 A Case Study Using Carboxylic Acid Chemical Categories to Evaluate Developmental Hazard to Species in Aquatic Environments

Aldrich's Flavors and Fragrance Catalog provides a list of carboxylic acids that are of commercial interest.[44] This list provides an excellent case study for the application of AOP-based chemical categories and read-across to assess developmental hazard to species developing in an aquatic environment.

Key intermediate event 4 in the AOP proposed above (Section 3.5.3) has been used as the basis for experimental assessment of developmental toxicity and structure-activity investigations with the frog *Xenopus*.[30,33] Specifically, this work focused on SCCAs of fewer than ten carbon atoms. Developmental toxicants were described by the developmental hazard index (DHI) defined by dividing 96-hr embryo mortality measured as the LC_{50} value by 96-hr embryo malformation measured by the EC_{50}.[33] *Xenopus* 96-hr LC_{50}, EC_{50} and DHI data are presented in Table 3.2. Chemicals with small DHI elicit malformations at concentrations near those which cause death. Since cessation of general cell functions near death can cause reduced embryo length and other generic malformations including oedema and abnormal gut coiling, SCCAs with low DHI values (DHI < 4.0) were not considered developmental hazards. Conversely, SCCAs eliciting a DHI of > 8.0 were considered strong developmental hazards. Acids with DHI values between 4.0 and 8.0, were considered weak developmental hazards.

Additional WoE supporting key intermediate event 4 in the AOP is provided from the *ex vivo* rat embryo data of Brown *et al.*[45] Briefly, 9.5-day conceptuses were exposed for 48 hours and, based on yolk-sac diameter, crown-rump length, number of somite pairs and overall morphology, a potency index (PI) from 1 (the most severe) to 6 (the least severe) was employed. Rat *ex vivo* PI data are presented in Table 3.3. While the DHI and PI values are not directly correlated there is the same general tread with both data sets. Structural alerts developed from these data confirm that compounds containing at least four, but fewer than six, carbon atoms in the chain with a tetrahedral α-carbon atom bound to a free carboxylic-group, one or two alkyl groups, and a hydrogen

Table 3.2 Toxicological data for 31 SCCAs measured with the 96-hr Frog Embryo Teratogenesis Assay-*Xenopus* (FETAX).

ID	Name	SMILES	MWt^a	LC_{50} mg/L	EC_{50} mg/L	LC_{50} mmol/L	ED_{50} mmol	DHI
	Alkyl-Straight Chain Saturated							
1	Formic acid	O=CO	46.03	4252	3581	92.37	77.80	1.2
2	Acetic Acid	O=C(O)C	60.05	6265	4418	104.33	73.57	1.4
3	Propionic acid	O=C(O)CC	74.08	7345	898	99.15	12.12	8.2
4	Butanoic acid	O=C(O)CCC	88.11	3559	422	40.39	4.79	8.4
5	Pentanoic acid	O=C(O)CCCC	102.13	2729	206	26.72	2.02	13.2
6	Hexanoic acid	O=C(O)CCCCC	116.16	1109	89.1	9.55	0.77	12.4
7	Heptanoic acid	O=C(O)CCCCCC	130.19	318.6	51.3	2.45	0.39	6.2
8	Octanoic acid	O=C(O)CCCCCCC	144.21	127.1	28.1	0.88	0.19	4.5
9	Nonanoic acid	O=C(O)CCCCCCCC	158.24	32.7	6.5	0.21	0.04	5.0
10	Decanoic acid	O=C(O)CCCCCCCCC	172.27	24	7.5	0.14	0.04	3.2
11	Undecanoic acid	O=C(O)CCCCCCCCCC	186.30	22.5	12.6	0.12	0.07	1.8
12	Dodecanoic acid	O=C(O)CCCCCCCCCCC	200.32	32.1	21.3	0.16	0.11	1.5
	Alkyl-Branched Chain Saturated							
13	2-Methylpropanoic acid	O=C(O)C(C)C	88.11	7925	2567	89.94	29.13	3.1
14	2-Methylbutanoic acid	O=C(O)C(CC)C	102.13	6591	1742	64.54	17.06	3.8
15	3-Methylbutanoic acid	O=C(O)CC(C)C	102.13	7010	531	68.64	5.20	13.2
16	2-Ethylbutanoic acid	O=C(O)C(CC)CC	116.16	7358	2011	63.34	17.31	3.7
17	2-Methyl pentanoic acid	O=C(O)C(C)CCC	116.16	2094	151	18.03	1.30	13.9
18	3-Methyl pentanoic acid	O=C(O)C(CC)C	116.16	3088	346	26.58	2.98	8.9
19	2-Propylpentanoic acid	O=C(O)C(CCC)CCC	144.21	852	30.5	5.91	0.21	27.9
20	2-Ethylhexanoic acid	O=C(O)C(CCCC)CC	144.21	646	47.3	4.48	0.33	13.7
	Alkyl-Straight Chain Unsaturated							
21	2-Butenoic acid	O=C(O)C=CC	86.09	4075	1190	47.33	13.82	3.4
22	trans-Pent-2-enoic acid	O=C(O)C=CCC	100.12	3184	472	31.80	4.71	6.7
23	4-Pentenoic acid	O=C(O)CC=C	100.12	2554	437	25.51	4.36	5.8
24	trans-Hex-2-enoic acid	O=C(O)C=CCCC	114.14	822	166	7.20	1.45	5.0

Table 3.2 (*Continued*)

ID	Name	SMILES	MWt^a	LC_{50} mg/L	EC_{50} mg/L	LC_{50} mmol/L	ED_{50} mmol	DH1
25	trans-Hex-3-enoic acid	O=C(O)CC=CCC	114.14	1501	306	13.15	2.68	4.9
	Alkyl Cyclic Aromatic							
26	Benzoic acid	O=C(O)c(cccc1)c1	122.12	1291	433	10.57	3.55	3.0
27	Phenylacetic acid	O=C(O)Cc(cccc1)c1	136.15	1274	802	9.36	5.89	1.6
28	2-Phenylpropionic acid	O=C(O)C(c(cccc1)c1)C	150.18	761	352	5.07	2.34	2.2
29	3-Phenylpropionic acid	O=C(O)CCc(cccc1)c1	150.18	729	123	4.85	0.82	5.9
30	2-Phenylbutanoic acid	O=C(O)C(c(cccc1)c1)CC	164.20	1368	563	8.33	3.43	2.4
31	3-Phenylbutanoic acid	O=C(O)CC(c(cccc1)c1)C	164.20	1124	241	6.85	1.47	4.7

^{a}Molecular Weight.

Table 3.3 Rat *ex vivo* derived Potency Indices (PI) for 18 SCCAs.

ID	Name	SMILES	MWt	PI
	Alkyl-Straight Chain Saturated			
1	Acetic Acid	O=C(O)CC	60.05	6
2	Propionic acid	O=C(O)CCC	74.08	2
3	Butanoic acid	O=C(O)CCCC	88.11	1
4	Pentanoic acid	O=C(O)CCCCC	102.13	2
5	Hexanoic acid	O=C(O)CCCCCC	116.16	3
6	Heptanoic acid	O=C(O)CCCCCCC	130.19	3
7	Octanoic acid	O=C(O)CCCCCCCC	144.21	4
	Alkyl-Branched Chain Saturated			
8	2-Methylpropionic acid	O=C(O)C(C)C	88.11	4
9	2-Methylbutanoic acid	O=C(O)C(CC)C	102.13	5
10	3-Methylbutanoic acid	O=C(O)CC(C)C	102.13	2
11	2-Ethylbutanoic acid	O=C(O)C(CC)CC	116.16	5
12	2-Methylpentanoic acid	O=C(O)C(C)CCC	116.16	5
13	2-Propylpentanoic acid	O=C(O)C(CCC)CCC	144.21	2
14	3-Methylpentanoic acid	O=C(O)C(CCC)C	116.16	4
15	2-Methyhexanoic acid	O=C(O)C(CCCC)C	130.19	5
16	2-Ethylhexanoic acid	O=C(O)C(CCCC)CC	144.21	3
	Alkyl-Straight Chain Unsaturated			
17	trans-Pent-2-enoic acid	O=C(O)C=CCC	100.12	6
18	4-Pentenoic acid	O=C(O)CCC=C	100.12	2

atom are likely to be developmental toxicants (Figure 3.3). In addition, there should be no double bond at the α- or β-carbon atoms and the alkyl groups must be larger than a methyl group. These structural requirements are similar to those described by Di Carlos and Narotsky *et al.* for teratogenesis in the mouse.[40,41] The major exception may be for the straight-chain SCCAs which elicit malformation *in vitro*. However, these compounds are unlikely to be teratogenic *in vivo* as they are predisposed to maternal metabolism via

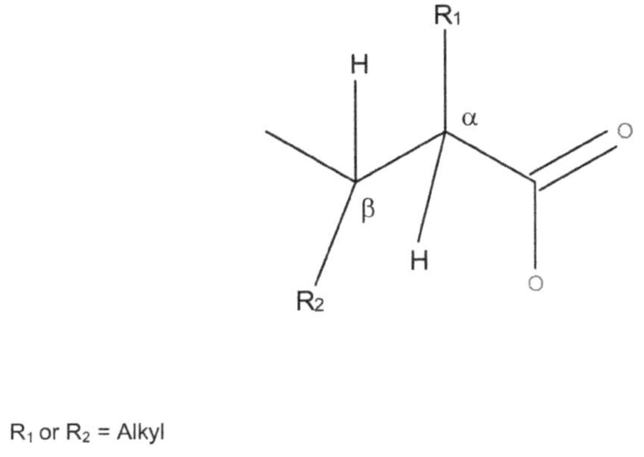

R_1 or R_2 = Alkyl

Figure 3.3 Structural alert identified from experimental data for SCCAs.

β-oxidation. As a result they are unlikely to persist long enough to cross the placenta and accumulate in the embryo *in utero*.[33,41,45]

The 39 carboxylic acids considered in this case study are listed in Table 3.4. Of these acids 28 have been evaluated for their DHI and 18 have been evaluated for their PI. The data gaps for the untested acids were filled by read-across based on those that had been tested, albeit with restrictions to sub-sets categories of acids. The results in Table 3.4 reveal 27 of the 39 acids in the case study inventory are ranked as non-developmental toxicants to aquatic species; these are acids that are shorter than three carbons or greater than eight carbons. Eight other acids are ranked as strong developmental hazards. These acids are typically four to six carbons in size. Lastly, the remaining four acids were ranked as weak developmental hazards.

Twelve untested acids, numbers 13, 14, 15, 23, 24, 25, 28, 29, 30, 31, 32 and 33 were all predicted to be non-developmental toxicants by read-across. In these cases, good experimental evidence for key intermediate event 4 is presented. Based on structural similarities, a category to fill the data gap for 9-decenoic acid has been formed (Figure 3.4). All six category members (without the target chemical) are longer (seven and more carbon atoms) straight chain saturated acids with experimental data showing them to be non-developmental toxicants, with the exception of the weak developmental toxicant — heptanoic acid. Although the target chemical is unsaturated, the double bond is far from the α and β carbons, and hence will not affect activity. The target acid, therefore, has been predicted to be a non-developmental toxicant based on the experimental results for the six acids. The read-across for acids 13, 14, 15, 30 and 31 has a higher uncertainty as it is based on extrapolation. The read-across for acids 34, 35 and 36 also has a higher level of uncertainty as they are based on a single analogue. Due to conflicting results, there is greater uncertainty associated with the prediction of acid 20, 2-methylpentanoic acid; similarly, there is less certainty in the prediction of acid 21, 3-methylpentanoic acid.

Confidence in a prediction of non-developmental toxicant would be strengthened by the addition of some experimental data for the other key events of the AOP. This reflects the reality that a false negative for non-toxic predictions has greater consequence that a false negative prediction for a toxic prediction.

Of greater concern are 4-methylpentanoic acid, 3-methyl-4-pentenoic acid, acids 22 and 27, respectively. In both of these cases, there is experimental evidence for highly similar analogues that suggest these two acids have considerable developmental hazard to species that develop in an aquatic environment. Clearly, further testing is needed to clarify the status of these acids as developmental toxicants.

In the case study present, an AOP has been used to provide the transparent mechanistic understanding as to why some carboxylic acids are developmental hazards and others are not. We used pre-existing, non-standard developmental toxicity studies to develop the AOP and *ex vivo* data and corresponding structural alert to justify the read-across to untested analogue(s). In each

Table 3.4 39 case study SCCAs with read-across predictions for untested chemicals.

ID	Name	SMILES	DHI	PI	Developmental Hazard Prediction
Alkyl-Straight Chain Saturated					
1	Formic acid	O=CO	1.2	–	non-developmental hazard supported by data for acid 2
2	Acetic Acid	O=C(O)C	1.4	6	non-developmental hazard supported by data for acid 1
3	Propionic acid	O=C(O)CC	8.2	2	strong developmental hazard; supported by data for acids 4, 5, 6, 26
4	Butanoic acid	O=C(O)CCC	8.4	1	strong developmental hazard; supported by data for acids 3, 5, 6, 26
5	Pentanoic acid	O=C(O)CCCC	13.2	2	strong developmental hazard supported by data for acids 3, 4, 6, 26
6	Hexanoic acid	O=C(O)CCCCC	12.4	3	strong developmental hazard supported by data for acids 3, 4, 5, 26
7	Heptanoic acid	O=C(O)CCCCCC	6.2	3	weak developmental toxicant; supported by trend for acids 5–12
8	Octanoic acid	O=C(O)CCCCCCC	4.5	4	non-developmental hazard; supported by trend for acids 7–12
9	Nonanoic acid	O=C(O)CCCCCCCC	5.0		non-developmental hazard; supported by trend for acids 7–12
10	Decanoic acid	CCCCCCCCC(=O)O	3.2		non-developmental hazard; supported by trend for acids 7–12
11	Undecanoic acid	O=C(O)CCCCCCCCCC	1.8		non-developmental hazard; supported by trend for acids 7–12
12	Dodecanoic acid	O=C(O)CCCCCCCCCCC	1.5		non-developmental hazard; supported by trend for acids 7–12
13	Tetradecanoic acid	O=C(O)CCCCCCCCCCCCC			non-developmental hazard; read-across from acid 12; supported by trend for acids 7–12
14	Pentadecanoic acid	O=C(O)CCCCCCCCCCCCCC			non-developmental hazard; read-across from acid 12; supported by trend for acids 7–12
15	Heptadecdecanoic acid	O=C(O)CCCCCCCCCCCCCCCC			non-developmental hazard; read-across from acid 12; supported by trend for acids 7–12
Alkyl-Branched Chain Saturated					
16	2-Methylpropionic acid	O=C(O)C(C)C	3.1	4	non-developmental hazard; supported by data for acids 17 and 19
17	2-Methylbutanoic acid	O=C(O)C(CC)C	3.8	5	non-developmental hazard; supported by data for acids16 and 19
18	3-Methylbutanoic acid	O=C(O)CC(C)C	13.2	2	strong developmental hazard; supported by data for 3-methyl pentanoic acid
19	2-Ethylbutanoic acid	O=C(O)C(CC)CC	3.7	5	non-developmental hazard; supported by data for acids 16, and 17
20	2-Methylpentanoic acid	O=C(O)C(C)CCC	13.9	5	strong developmental toxicant; mixed results lowers reliability of prediction

Table 3.4 (*Continued*)

ID	Name	SMILES	DHI	PI	Developmental Hazard Prediction
21	3-Methylpentanoic acid	O=C(O)CC(CCC)C	8.9	4	weak developmental toxicant; results for acid 18 lowers reliability of prediction
22	4-Methylpentanoic acid	O=C(O)CCC(C)C			strong developmental hazard; read-across from acids 5 and 6
23	2-Methylheptanoic acid	O=C(O)C(CCCCC)C			non-developmental hazard read-across from acids 16, and 17; supported by trend for acids 7–12
24	4-Methyloctanoic acid	O=C(O)CCC(CCCC)C			non-developmental hazard; read-across from acid 9; supported by trend for acids 7–12
25	4-Methylnonanoic acid	O=C(O)CCC(CCCCC)C			non-developmental hazard read-across from acid 10; supported by trend for acids 7–12
Alkyl-Straight Chain Unsaturated					
26	4-Pentenoic acid	O=C(O)CCC=C	5.8	2	weak developmental hazard; supported by data for trans-3-hexenoic acid; results for acids 5 and 6 lowers reliability
27	3-Methyl-4-pentenoic acid	O=C(CC(C=C)C)O			strong developmental hazard; read-across from acid 20
28	9-Decenoic acid	OC(=O)CCCCCCCC=C			non-developmental hazard; read-across from acid 10; supported by trend for acids 7–12
29	10-Undecylenic acid	O=C(O)CCCCCCCCC=C			non-developmental hazard; read-across from acid 11; supported by trend for acids 7–12
30	9-Octadecenoic acid	O=C(O)CCCCCCCC=CCCCCCCCC			non-developmental hazard; read-across from acid 12; supported by trend for acids 7–12
31	9,12-Octadecadieneoic acid	O=C(O)CCCCCCCC=CCC=CCCCCC			non-developmental hazard; read-across from acid 12; supported by trend for acids 7–12
Alkyl-Branched Chain Unsaturated					
32	3-Methyl-2-butenoic acid	O=C(O)C=C(C)C			non-developmental toxicant; read-across from 2-butenoic acid
33	3,7-Dimethyl-6-octenoic acid	O=C(O)CCC(CCC=C(C)C)C			non-developmental toxicant; read-across from acids 8, 9, 10
Alkyl Cyclic Saturated					

Table 3.4 (Continued)

ID	Name	SMILES	DH1	PI	Developmental Hazard Prediction
34	Cyclohexanecarboxylic acid	O=C(O)C(CCCC1)C1			non-developmental hazard; read-across from acid 38
35	Cyclohexaneacetic acid	O=C(O)CC(CCCC1)C1			non-developmental hazard; read-across from acid 38
36	1-Cyclopentene-2-acetic acid	O=C(O)C(C=CCC1)C1			non-developmental hazard; read-across from acid 38
Alkyl Cyclic Aromatic					
37	Benzoic acid	O=C(O)c(cccc1)c1	3		non-developmental hazard
38	Phenylacetic acid	O=C(O)Cc(cccc1)c1	1.6		non-developmental hazard
39	3-Phenylpropionic acid	O=C(O)CCc(cccc1)c1	5.9		weak developmental toxicant; supported by data for acid 7

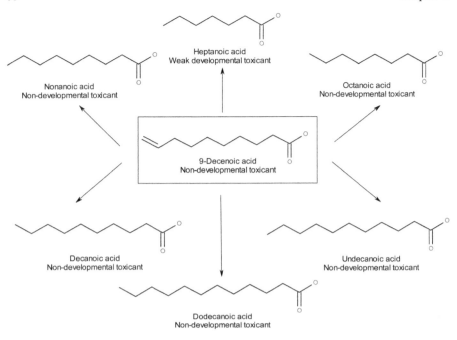

Figure 3.4 Chemical category developed for the target chemical 9-decenoic acid allowing for the prediction of developmental toxicity hazard.

scenario, the data used were for key event 4, the inhibition of neural tube elongation and closing, which reflects convergent extension in early embryos. Both data sets used in the WoE are causally linked to the initial upstream event hypothesised as the inhibition of histone deacetylase. In principle, each data set could be considered sufficient to characterise the developmental toxicity hazard. However, by using them in tandem one can use WoE to address uncertainty in the prediction.

Because of the nature of the available data, the assessments were limited to simple carboxylic acids as developmental toxicants to organisms developing in aquatic environments. However, this AOP, along with chemical categories, read-across and WoE approaches could be adapted to assess endpoints relevant to humans for selected carboxylic acids and related compounds. Specifically, candidate chemicals could be assessed for release of acetyl groups and/or accumulation of β-catenin. Subsequently, a subset of chemicals could be assessed for cell adhesion and/or cell motility or convergent extension in early embryos either tested *ex vivo* in mammalian embryos of using lower vertebrates such as fish or amphibians. Finally, the data gap(s) could be filled from a few chemicals (*e.g.* valproic acid, 2-ethylhexanoic acid) measured *in vivo* with *in utero* exposed mammals for the apical adverse effect. Particular attention will need to be paid to straight-chain SCCAs which elicit malformation *in vitro*, but it is unlikely there would be teratogenicity *in vivo*.

3.6 Conclusions

The AOP methodology is an approach which provides a framework to collect, organise and evaluate relevant mechanistic data from the various levels of biological complexity. In this context, an AOP is developed as a sequence of key events starting from the chemical-biological interaction at the molecular level through the cascade of biological intermediate events and ending on the *in vivo* outcome. Using such organised knowledge in category formation enables grouping of chemicals based on their chemical as well as biological properties. The possibility to categorise chemicals based on their similarity in toxicological behaviour is the most important advantage of the AOP approach in chemical grouping. Compounds gathered in the same category will share not only an MIE but also one or more key events resulting from that chemical-biological interaction. Such a category will provide stronger confidence that all members elicit the same final adverse effect. Therefore, the AOPs would allow for the transition from a chemical-by-chemical approach to a chemical-category based approach in the assessment of chemicals. The usage of an AOP developed for one compound to assess toxicity for other similar chemicals is highly desirable for risk assessment because it reduces cost, effort, time, and the use of test animals.

Acknowledgements

The funding from the European Community's 7th Framework Program (FP7/2007–2013) COSMOS Project under grant agreement no. 266835 and from Cosmetics Europe is gratefully acknowledged. Funding from the European Chemicals Agency (Service Contract No. ECHA/2008/20/ECA.203) is also gratefully acknowledged.

References

1. T. W. Schultz, Adverse outcome pathways: A way of linking chemical structure to *in vivo* toxicological hazards in *In Silico Toxicology: Principles and Applications*, ed. M. T. D. Cronin and J. C. Madden, Royal Society of Chemistry, Cambridge, 2005, p. 346.
2. G. T. Ankley, R. S. Bennett, R. J. Erickson, D. J. Hoff, M. W. Hornung, R. D. Johnson, D. R. Mount, J. W. Nichols, C. L. Russom, P. K. Schmieder, J. A. Serrano, J. E. Tietge and D. L. Villeneuve, Adverse Outcome Pathways: A conceptual framework to support ecotoxicology research and risk assessment, *Environ. Toxicol. Chem.* 2010, **29**, 730.
3. H. J. Clewell, P. R. Gentry, J. M. Gearhart, B. C. Allen and M. E. Andersen, Considering pharmacokinetic and mechanistic information in cancer risk assessments for environmental contaminants: examples with vinyl chloride and trichloroethylene, *Chemosphere*, 1995, **31**, 2561.

4. D. L. Shuey, R. W. Setzer, C. Lau, R. M. Zucker, K. H. Elstein, M. G. Narotsky, R. J. Kavlock and J. M. Rogers, Biological modeling of 5-fluorouracil developmental toxicity, *Toxicology*, 1995, **102**, 207.
5. C. Sonich-Mullin, R. Fielder, J. Wiltse, K. Baetcke, J. Dempsey, P. Fenner-Crisp, D. Grant, M. Hartley, A. Knaap, D. Kroese, I. Mangelsdorf, E. Meek, J. M. Rice and M. Younes, International Programme on Chemical Safety. IPCS conceptual framework for evaluating a mode of action for chemical carcinogenesis, *Regul. Toxicol. Pharmacol.*, 2001, **34**, 146.
6. National Research Council (NRC), *Toxicity Testing in the 21st Century: A Vision and a Strategy*, Washington, DC:The National Academies Press, 2007.
7. The Organisation for Economic Co-operation and Development (OECD) (2011) *Report of the Workshop on Using Mechanistic Information in Forming Chemical Categories.* OECD Environment, Health and Safety Publications Series on Testing and Assessment No. 138. ENV/JM/MONO(2011)8.
8. The Organisation for Economic Co-operation and Development (OECD) (2012) *Proposal for a Template, and Guidance on Developing and Assessing the Completeness of Adverse Outcome Pathways.* http://www.oecd.org/env/ehs/testing/49963554.pdf
9. T. W. Schultz, B. Diderich and S. Enoch, The OECD Adverse Outcome Pathway Approach a case study for skin sensitisation in *Alterative Testing Strategies Progress Report 2011 & AXLR8-2 Workshop Report on a Roadmap to Innovative Toxicity Testing*, ed. T. Seidle and H. Spielmann, 2011, p. 288.
10. D. A. Keller, D. R. Juberg, C. Catlin, W. H. Farland, W. G. Hess, D. C. Wolf and N. G. Doerrer, Identification and characterization of adverse effects in 21st century toxicology, *Toxicol. Sci.* 2012, **126**, 291.
11. S. J. Enoch, C. M. Ellison, T. W. Schultz and M. T. D. Cronin, A review of the electrophilic reaction chemistry involved in covalent protein binding relevant to toxicity, *Crit. Rev. Toxicol.*, 2011. **41**, 783.
12. I. Tsakovska, I. Pajeva, P. Alov and A. Worth, Recent advances in the molecular modeling of estrogen receptor-mediated toxicity, in *Advances in Protein Chemistry and Structural Biology*, ed. C. Christov, Academic Press, Burlington, 2011, p. 217.
13. J. Hooyberghs, E. Schoeters, N. Lambrechts, I. Nelissen, H. Witters, G. Schoeters, G. and R. Van Den Heuvel, A cell-based *in vitro* alternative to identify skin sensitizers by gene expression, *Toxicol. Appl. Pharmacol.*, 2008, **231**, 103.
14. A. B. Hill, The environmental and diseases: Association or causation?, *Proc. Roy. Soc. Med.*, 1965, **58**, 295.
15. D. C. Volz, S. Belanger, M. Embry, S. Padilla, H. Sanderson, K. Schirmer, S. Scholz and D. Villeneuve, Adverse outcome pathways during early

fish development: a conceptual framework for identification of chemical screening and prioritization strategies, *Toxicol. Sci.*, 2011, **123**, 349.
16. K. H. Watanabe, M. E. Andersen, N. Basu, M. J. Caravan 3rd, K. M. Crofton, K. A. King, C. Sunol, E. Tiffany-Castiglioni and I. R. Schultz, Defining and modeling known adverse outcome pathways: domoic acid and neuronal signaling as a case study, *Environ. Toxicol. Chem.*, 2011, **30**, 9.
17. K. L. Yozzo, S. P. McGee and D. C. Volz, Adverse outcome pathways during zebrafish embryogenesis: A case study with paraoxon, *Aquat. Toxicol.*, 2013, **126**, 346.
18. The Organisation for Economic Co-operation and Development (OECD) (2012) *The Adverse Outcome Pathway for Skin Sensitization Initiated by Covalent Binding to Proteins. Part 1. Scientific Evidence.* Part 1 scientific evidence. OECD Environment, Health and Safety Publications Series on Testing and Assessment No. 168. ENV/JM/MONO(2012)10/PART1.
19. The Organisation for Economic Co-operation and Development (OECD) (2012) *The Adverse Outcome Pathway for Skin Sensitization Initiated by Covalent Binding to Proteins. Part 2. Use of AOP to Develop Chemical Categories and Integrated Assessment and Testing Approaches.* OECD Environment, Health and Safety Publications Series on Testing and Assessment No. 169. ENV/JM/MONO(2012)10/PART2.
20. http://www.effectopedia.org/.
21. International Programme on Chemical Safety (IPCS), *Mode of Action Framework*, World Health Organization, Geneva, 2008.
22. The Organisation for Economic Co-operation and Development (OECD) (2011) *Provision of Knowledge and Information: Hazard Assessment: Follow-up from the OECD Workshop on Using Mechanistic Information in Forming Chemical Categories: a Long-term Vision for the (Q)SAR Project.* ENV/JM(2011)6.
23. The OECD QSAR Application Toolbox. (available from: http://www.qsartoolbox.org/).
24. Y. Sakuratani, H. Q. Zhang, S. Nishikawa, K. Yamazaki, T. Yamada, J. Yamada and M. Hayashi, Categorization of nitrobenzenes for repeated dose toxicity based on adverse outcome pathways, *SAR QSAR Environ. Res.*, 2013, **24**, 35.
25. R. H. Finnell, J. Gelineau-van Waes, J. D. Eudy and T. H. Rosenquist, Molecular basis of environmentally induced birth defects, *Annu. Rev. Pharmacol. Toxicol.*, 2002, **42**, 181.
26. I. Naruse, M. D. Collins and W. J. Jr. Scott, Strain differences in the teratogenicity induced by sodium valproate in cultured mouse embryos, *Teratology*, 1988, **38**, 87.
27. J. Behrens and M. Kühl, Wnt signaling transduction pathways: An Overview, in *Wnt Signaling in Development*, ed. M. Kühl, Kluwer Academic/Plenum, Publishers, New York, 2003, p. 1.

28. N. Khan, M. Jeffers, S. Kumaar, C. Hackett, F. Boldog, N. Khramtsov, X. Qian, E. Mills, S. C. Berghs, N. Carey, P. W. Finn, L. S. Collins, A. Tumber, J. W. Ritchie, P. B. Jensen, H. S. Lichenstein and M. Sehested, Determination of the class and isoform selectivity of small-molecule histone deacetylase inhibitors, *Biochem. J.*, 2008, **409**, 581.
29. J. M. Mehnert and W. K. Kelly, Histone deacetylase inhibitors: Biology and mechanism of action, *Cancer J*, 2007, **13**, 23.
30. C. J. Phiel, F. Zhang, E. Y. Huang, M. G. Guenther, M. A. Lazar and P. S. Klein, Histone deacetylase is a direct target of valproic acid, a potent anticonvulsant, mood stabilizer, and teratogen, *J. Biol. Chem.*, 2001, **276**, 36734.
31. J. Wiltse, Mode of action: Inhibition of histone deacetylase, altering Wnt-dependnent gene expression and regulation of beta-catenin-Developmental effects of valproic acid, *Crit. Rev Toxicol.*, 2005, **35**, 727.
32. L. C. Fuller, S. K. Cornelius, C. W. Murphy and D. J. Wiens, Neural crest cell motility in valporic acid, *Reprod. Toxicol.*, 2002, **16**, 825.
33. D. A. Dawson, T. W. Schultz and R. S. Hunter, Developmental toxicity of carboxylic acids to *Xenopus* embryos: A quantitative structure-activity relationship and computer-automated structure evaluation, *Teratogen. Carcinogen. Mutagen.*, 1996, **16**, 109.
34. J. B. Wallingford and R. M. Harland, Neural tube closure requires disheveled-dependent convergent extension of the midline, *Development*, 2002, **129**, 5815.
35. C. Yost, M. Tones, J. R. Miller, E. Huang, D. Kimelman and R. T. Moon, The axis-inducing activity, stability, and subcellular distribution of beta-catenin is regulated in *Xenopus* embryos by glycogen synthase kinase 3, *Genes Devel.*, 1996, **10**, 1443.
36. C. Kliecker and C. Niehrs, A morphogen gradient of Wnt/β-catenin signaling regulates anteroposterior neural patterning in *Xenopus*, *Development*, 2001, **128**, 4189.
37. J. B. Wang, N. S. Hamblet, S. Mark, M. E. Dickinson, B. C. Brinkman, N. Segil, S. E. Fraser, P. Chen, J. B. Wallingford and A. Wynshaw-Boris, Dishevelled genes mediate a conserved mammalian PCP pathway to regulate convergent extension during neurulation, *Development*, 2006, **133**, 1767.
38. E. Robert and P. Guibaud, Maternal valproic acid and congenital neural tube defects, *Lancet*, 1982, **2**, 937.
39. E. Robert and F. Rosa, Valproate and birth defects, *Lancet*, 1983, **2**, 1.
40. E. J. DeCarlos, Structure-activity relationships (SAR) and structure metabolism relationships (SMR) affecting the teratogenicity of carboxylic acids, *Drug Metab. Rev.*, 1990, **22**, 411.
41. M. G. Narotsky, E. Z. Francis and R. J. Kavlock, Developmental toxicity and structure-activity relationship of aliphatic acids, including dose-response assessment of valproic acid in mice and rats, *Fundam. Appl. Toxicol.*, 1994, **22**, 251.

42. B. Heltweg, J. Trapp and M. Jung, *In vitro* assays for the determination of histone deacetylase activity, *Methods*, 2005, **36**, 332.
43. T. Ciossek, H. Julius, H. Wieland, T. Maier and T. Beckers, A homogenous cellular histone deacetylase assay suitable for compound profiling and robotic screening, *Anal. Biochem.*, 2007, **371**, 72.
44. http://www.sigmaaldrich.com/chemistry/chemistry-products.html?TablePage=12916829
45. N. A. Brown, M. E. Coakley and D. O. Clarke, Structure-teratogenicity relationships of valproic acid congeners in whole-embyro culture, in *Approaches to Elucidate Mechanism of Teratogenesis*, ed. F. Welsch, Hemisphere, Washington, 1987, p. 17.

CHAPTER 4
Tools for Grouping Chemicals and Forming Categories

J. C. MADDEN

School of Pharmacy and Chemistry, Liverpool John Moores University, Byrom Street, Liverpool, L3 3AF, England
E-mail: j.madden@ljmu.ac.uk

4.1 Introduction

Advances in computational chemistry and high-throughput screening have resulted in a vast amount of information becoming available on chemicals, including data on chemical reactivity, structural, physico-chemical and toxicological properties. However, there are still many data gaps that need to be addressed, particularly with respect to the toxicological profile of chemicals. Grouping methods allow knowledge concerning chemicals with known properties to be used to infer information about those compounds for which the properties are unknown; this is the process of read-across. This relies upon compounds being organised into rationally-based, clearly identifiable groups that share a particular, relevant property or properties. The rationale for forming such groups must be fully transparent and justifiable if the resulting read-across predictions are to be acceptable. Similarities in structure, size, 3-dimensional shape, physico-chemical properties, presence of functional groups or identified structural alerts, chemical reactivity, mechanism or mode of action and biological activity are all potential criteria on which to form a group. This chapter focuses on currently available tools that can be used for grouping, including bespoke tools (such as the OECD QSAR Toolbox) and other tools, designed with broader applications, that can potentially be used to identify similar compounds (such as Toxtree or ChemSpider).

4.2 Reasons for Grouping Compounds

The criteria on which a category may be built depend upon the information available and the purpose of the grouping exercise; different scenarios may be encountered, for example:

(i) When investigating compounds associated with a specific toxicological endpoint (such as mutagenicity) it may be possible to group compounds into categories based on relevant properties *e.g.* chemical reactivity, presence of specific structural features *etc.* For example, knowledge of mechanistic organic chemistry can be used to identify groups of compounds that may be associated with a given toxicity. Identifying the characteristics of compounds with similar activity provides fundamental knowledge for the development of profilers that can be used subsequently to identify other molecules with the same characteristics and therefore potentially the same activity.

(ii) A second scenario involves the investigation of the potential toxicity of a given target chemical, for which no toxicity data are available, by using data from similar compounds. In this case the target chemical can be assessed using existing profilers that are associated with a given toxicity. Compounds, similar to the target chemical, but for which toxicity data are available, are identified and used to form a category from which the activity of the unknown is inferred *i.e.* the category is formed for the purpose of read-across. Read-across may be performed qualitatively (*e.g.* to give a prediction of active/inactive); semi-quantitatively (*e.g.* inactive, weak, medium or high potency) or quantitatively (*e.g.* predicting defined potency values).

(iii) The third scenario is one which is becoming increasingly relevant given the current drive to obtain large amounts of information using high-throughput screening (HTS) methods. Initiatives such as the ToxCast and Tox21 programs (http://epa.gov/ncct/Tox21/) have, or are in the process of, screening large numbers of compounds through hundreds of high-throughput screens. This has resulted in a vast amount of information (including *in vitro* and -omics data) being generated for these compounds. Such volumes of information are difficult to analyse using traditional methods; what is required is a method to rationalise these high volume data to investigate how (mammalian) toxicity may be elicited. Results from HTS can be used to group compounds into categories based on their *in vitro* or -omics profiles. If toxicity data are available for some chemicals within a given group this can be used to provide insight into potential mechanisms of toxicity, or a toxicity profile, for other members of the group.

Of the three scenarios given above it is the second that is the focus here *i.e.* grouping compounds together in a rational manner to obtain read-across predictions for toxicity. In certain cases grouping may be carried out intuitively by those with appropriate expertise, but there is also an array of software applications that also may be applied. As discussed in Chapter 2, there are different approaches that may be used for grouping chemicals:

(i) analogue approaches, where compounds may be grouped together based on the presence of a particular functional group with a small variation in structure *e.g.* carbon chain length;
(ii) mechanistic approaches, whereby compounds that may elicit the same molecular initiating event are grouped together and;
(iii) chemoinformatics based approaches that rely on structural information concerning the compounds of interest.

The aim of this chapter is to introduce different tools that are available to group chemicals together based on different approaches identified. Within one chapter it is only possible to give a brief overview of the capabilities and potential application of these tools. Chapter 6 provides case studies as examples of where some of the tools have been applied; in all cases references are given where further information may be obtained.

4.3 The OECD QSAR Toolbox

One of the most important tools for grouping and read-across is the OECD QSAR Toolbox, hereafter referred to in this chapter as the Toolbox. The development of the Toolbox was co-ordinated by the Organisation for Economic Co-operation and Development (OECD) and the work undertaken in the Laboratory of Mathematical Chemistry at the University "Prof. Assen Zlatarov", Bourgas, Bulgaria under the leadership of Professor Ovanes Mekenyan. The software was developed in collaboration with the European Chemicals Agency (ECHA). The Toolbox (for its history see Section 1.3) was designed specifically for the purpose of category formation and read-across to fill gaps in data needed for safety/hazard assessment of chemicals. It was specifically developed to be used by the chemical industry and other stakeholders to help with regulatory submissions where *in silico* tools were employed for predicting toxicity. The first version of the Toolbox was launched in March 2008; the most recent version (at time of writing) is version 3.1 which was released in January 2013. The Toolbox is freely downloadable, along with detailed user guides and supporting information (http://www.qsartoolbox.org). The software is promoted as providing (in certain cases) an alternative to animal testing to provide (eco)toxicity hazard information required for REACH submissions. Analogues of a target chemical can be identified; data for these can be retrieved from databases within the Toolbox and read-across or trend analysis can be used to fill the data gap. Other applications of the Toolbox include the ability to categorise chemical inventories by mechanism or mode of action using the profilers available within the software. Detailed reports are also generated by the software providing the documentation necessary to support a given prediction.

4.3.1 The Workflow of the Toolbox

An overview of the information (*e.g.* databases) and applications (*e.g.* description of profiling tools) available within the Toolbox are briefly described below. The "Getting Started" guide for the Toolbox available at http://oasis-lmc.org/products/software/toolbox/toolbox-support.aspx provides further details and practical guidance on using the functionalities within the Toolbox.[1] Additional information, such as descriptions of the profilers, is available as supporting information within the software itself.

The Toolbox is a bespoke tool for the grouping of compounds into rational, chemically and/or mechanistically justifiable categories. These categories can be built using structural or mechanistic features of chemicals that are relevant to the toxicological endpoint being investigated. Further sub-categorisation can then be performed as required using profilers, already defined within the Toolbox, or user-defined profilers to ensure that the members of the category fall within a clearly defined structural domain, representative of the target chemical. Read-across can then performed from those members of the category with known experimental data to those where data are lacking. A full report on the process can be generated in-keeping with regulatory requirements. The software follows a clear workflow, which is designed to mimic the manner in which an assessor would make a judgement on a chemical. The workflow is shown in Figure 4.1 and described in more detail below.

Step (1) of Figure 4.1 is the input of the chemical(s) of interest into the Toolbox. For a single chemical the input may be from a file or in the form of a Chemical Abstracts Service (CAS) registry number, name, chemical identifier (EC number or EINECS number) SMILES string, InChI, structure drawn using the editor within the software, or selected from one of the program's resident chemical databases or inventories. Multiple chemicals can also be entered as a batch using CAS numbers, SMILES strings or from files or databases in SDF, MOL, MOL2, RDF and XYZ format. Having entered the required structure into the program it is essential to check that it is the correct structure as subsequent functions are based on specific characteristics of the structure used. The Toolbox contains "quality checked" structures for over 300,000 chemicals ensuring the user has access to the "correct" structures for read-across.

One of the advantages of the Toolbox is that as an international collaborative project, many databases of toxicity and other data, as well as inventories have been donated to the project. Table 4.1 lists the databases and inventories present in version 3.1 of the Toolbox, however, it should be noted that these are updated regularly.

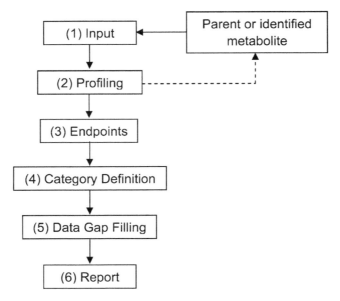

Figure 4.1 Workflow of the OECD QSAR Toolbox.

Step (2) is the profiling of the chemical(s) that has been entered. This is used to identify relevant structural features, or potential mechanisms of action of the chemical. There are five categories of profilers available within the Toolbox, these are shown in Table 4.2. (Information concerning these profilers was obtained from the supporting information within the software.) Alternatively users may define their own profilers. If the user is interested in a given endpoint that has a known associated mechanism, then the choice of profiler may be obvious. For example, an initiating step in skin sensitisation is the formation of a covalent bond between the chemical and a skin protein.[2] If skin sensitisation is the endpoint of interest then the chemical(s) can be profiled using protein binding alerts. Similarly, genotoxicity can be elicited by a covalent bond formation between a chemical and DNA, in this case using a DNA binding profiler is appropriate.[3] The profilers within the Toolbox have been built using expert knowledge of chemical interactions; examples of using these mechanistic profilers are given in Chapter 6. Chemicals can also be profiled according to specific endpoints, for example profiling based on structural alerts for mutagenicity or carcinogenicity. More general profilers are also available such as affiliation with a known database, or presence of particular chemical elements or organic functional groups.

It is also possible to obtain a metabolic profile for chemicals of interest. Table 4.3 shows the metabolic profilers that are available. Once metabolites have been identified for a given chemical the metabolites can then be profiled using any of the profilers outlined in Table 4.2. This provides useful

Table 4.1 Databases and inventories available in the OECD QSAR Toolbox version 3.1.

Database	Content
Aquatic ECETOC	Experimental results for aquatic toxicity
Aquatic Japan MoE	Experimental results on aquatic toxicity based on tests performed within the Japanese Existing Chemicals Programme
Aquatic OASIS	Experimental results for aquatic toxicity gathered from different sources
Aquatic US EPA ECOTOX	A comprehensive database which provides information on adverse effects of single chemical stressors to ecologically relevant aquatic species
Bacterial mutagenicity ISSSTY	ISSSTY database on *Salmonella typhirmurium* (AMES test)
Bioaccumulation Environment Canada	Experimental results on bioaccumulation in aquatic organisms
Bioaccumulation fish CEFIC-LRI	Experimental results on bioaccumulation values in fish
Bioconcentration NITE	Bioconcentration test data for existing chemicals under the Japanese Chemical Substances Control Law conducted by Ministry of Economy, Trade and Industry (METI)
Biodegradation in soil OASIS	Experimental results for ready biodegradation in soil
Biodegradation NITE	Biodegradation test data from METI
Biota-Sediment Accumulation Factor US EPA	BSAF is a dataset of approximately 20,000 biota-sediment accumulation factors
Carcinogenicity Potency Database CPDB	Experimental results for long term animal cancer studies
Carcinogenicity and Mutagenicity ISSCAN	This database includes experimental results for genotoxicity and carcinogenicity
Cell transformation assay ISSCTA	Results of four *in vitro* cell transformation assays for detection of chemical carcinogens
Chemical Reactivity Colipa	Chemical reactivity data relating to cysteine and/or lysine depletion and adduct formation
Dendritic cells COLIPA	Results of CD54 and CD86 expression (biomarkers linked to skin sensitisation)
Developmental toxicity ILSI	NOEL and LOEL data for different species and routes of administration including maternal and foetal effects
ECHA CHEM	Information on chemicals manufactured or imported into Europe, provided in registration dossiers submitted to the European Chemicals Agency (ECHA). Level of information available dependent on the chemical
ERBA OASIS	Data on Estrogen Receptor Binding Affinity (ERBA) expressed as relative binding affinities in comparison with the estradiol affinity
Experimental pKa	Experimental pKa values
Eye irritation ECETOC	Experimental results for rabbit eye irritation

Table 4.1 (*Continued*)

Database	Content
Genotoxicity OASIS	Experimental results for genotoxicity
GSH experimental RC50	Includes empirical abiotic thiol reactivity data expressed as the *in chemico* RC_{50} value for electrophiles
Hydrolysis rate constant OASIS	Experimental data for neutral hydrolysis
kM database Environment Canada	Estimated whole body *in vivo* metabolic transformation rate constants (kM) values for fish determined using measured laboratory bioconcentration factors and total elimination rate constants
Micronucleus ISSMIC	A curated database, containing critically-selected information on chemical compounds tested with the *in vivo* micronucleus mutagenicity assay in rodents
Micronucleus OASIS	The micronucleus database consists of 577 chemicals having *in vivo* bone marrow and peripheral blood MNT data
Munro non cancer EFSA	No Observed Effect Level (NOEL) and Lowest Observed Effect Level (LOEL) data from Munro dataset
Phys-Chem EPISUITE	This database includes experimental results on physical chemical properties as accessed from EPISUITE. This is an extract from the PHYSPROP database maintained at Syracuse Research Corporation
Repeated dose toxicity HESS	Contains information on repeated dose toxicity of 289 industrial chemicals
Rodent inhalation toxicity database	Experimental data from rat inhalation studies
Skin irritation	Includes Primary Skin Irritation Indices from skin irritation test from several sources
Skin sensitisation	Includes experimental results for skin sensitisation donated from various sources
Skin sensitisation ECETOC	Includes experimental results on skin and respiratory sensitisation
Terrestrial US EPA ECOTOX	ECOTOX is a comprehensive database, which provides information on adverse effects of single chemical stressors to ecologically relevant terrestrial species
Toxicity Japan MHLW	Ministry of Health Labour and Welfare Includes experimental results from single dose toxicity tests and mutagenicity tests performed under the Japanese Existing Chemicals Programme
ToxRefDB US EPA	Contains information on chronic developmental and reproductive studies and studies on cancer for more than 300 pesticides
Yeast estrogen assay database University of Tennessee USA	Organic compounds binding to the estrogen receptor can activate gene expression. Relative gene activation data available for 213 chemicals

Table 4.1 (*Continued*)

Inventory	Content
AICS	Australian Inventory of Chemical Substances: List of all chemicals in use in Australia 1977–1990 and newly assessed chemicals
Canada DSL	Canada's Domestic Substance List of chemicals manufactured, imported into or used in Canada at a commercial level
COSING	European Commission Database with information on cosmetic substances and ingredients
DSS-Tox	Distributed Structure-Searchable Toxicity Database Network from the US Environment Protection Agency
ECHA PR	European Chemicals Agency Pre-Registered Substances List
EINECS	European Inventory of Existing Chemical Substances
HPVC OECD	OECD list of High Production Volume Chemicals
METI Japan	Ministry of International Trade and Industry: Japanese Existing and New Chemical Substances List
REACH ECB	List of substances pre-registered under REACH (responsibility has now moved from the European Chemicals Bureau (ECB) to ECHA)
TSCA	List of chemical substances manufactured or processed in the US required under the Toxic Substances Control Act
US HPV Challenge Program	List of High Production Volume Chemicals within the US

information, for example in cases where the metabolites are responsible for eliciting a toxic effect rather than the parent chemical.

Profilers are continually being developed for the Toolbox and it is important to identify the most appropriate profiler (or profilers) for a given investigation based on knowledge of the endpoint of interest and, where possible, an understanding of the underlying mechanisms. Examples of where specific profilers have been chosen are given in case studies 1–4 in Chapter 6. It must be emphasised that the decision of which profiler to use may be a subjective choice and will require expertise not only in the use of the Toolbox but also in (mechanistic) toxicology.

Step (3) is to gather data on the endpoint(s). Data gathering can be performed on an individual, specific endpoint or on a range of endpoints selected by the user. Data trees are used to classify the endpoint data into increasingly specific categories. For example aquatic toxicity can be iteratively sub-classified into all fish > specific fish > endpoint > specified time point *etc*. Information can be gathered from the databases within the Toolbox. This includes all of the databases listed in Table 4.1 in addition to the "RepDoseTox Fraunhofer ITEM" database which includes repeat dose toxicity data for 615 chemicals in rats and mice. Alternatively the user can upload information from their own databases. Hence, this step identifies chemicals for which relevant endpoint data may be available. Once the suitability of the chemicals has been assessed for inclusion in the category these

Table 4.2 Profilers available in the OECD QSAR Toolbox version 3.1.

Profiler	Further information
Pre-defined Profilers	
Database affiliation	Affiliation with one or more of the databases indicated in Table 4.1
Inventory affiliation	Affiliation with one or more of the inventories indicated in Table 4.1
OECD High Production Volume (HPV) Chemical Categorie	Developed in list of high production volume chemicals (produced or imported at $>$ 1,000 tonnes per year)
Substance type	Identifies 5 substance types (i) discrete chemicals (ii) dissociating chemicals (iii) mixtures (iv) polymers (v) chemicals of no defined composition
US Environment Protection Agency (EPA) New Chemical Categories	Reproduces the original categories of the "Toxic Substances Control Act New Chemicals Program NCP)/Chemical Categories" (certain categories are excluded).
General Mechanistic Profilers	
BioHC half-life (Biowin)	Uses a fragment based approach to estimate quantitative biodegradation half-lives for hydrocarbons
Biodeg primary (Biowin 4)	Estimates time for primary biodegradation (*i.e.* to form initial metabolite) in a typical aquatic environment
Biodeg probability (Biowin 1)	A linear biodegradation probability calculated based on fragment values which is converted into ranges of "biodegrades fast" or "does not biodegrade"
Biodeg probability (Biowin 2)	A non-linear biodegradation probability calculated based on fragment values which is converted into ranges of "biodegrades fast" or "does not biodegrade"
Biodeg probability (Biowin 5)	Based on the Biowin 5 model in EPISUITE (property prediction program from the US EPA) for assessing biodegradability in the Japanese MITI biodegradation test. Uses a fragment constant approach, converts result into "Readily biodegradable" or "Not readily biodegradable"
Biodeg probability (Biowin 6)	Similar to Biowin 5 above, but based on Biowin 6 in EPISUITE
Biodeg probability (Biowin 7)	Based on Biowin 7 in EPISUITE, estimates probability of fast biodegradation under methanogenic anaerobic conditions, predictive of degradation in typical anaerobic digester. Indicates probability that compound "biodegrades fast" or "does not biodegrade fast"
Biodeg ultimate (Biowin 3)	Estimates time for complete biodegradation in a typical aquatic environment
DNA binding by OASIS v.1.1	Identifies structural requirements for mutagenicity, based on the AMES mutagenicity models in the TIMES system. Includes 49 categories relating to interaction with DNA. Definition of the alerts justified by their interaction mechanisms found in the literature

Table 4.2 (*Continued*)

Profiler	Further information
DNA binding by OECD	Structural alerts for the binding of organic chemicals to DNA. Includes 87 categories relating to 6 broad organic chemistry mechanisms
DPRA cysteine peptide depletion	Direct Peptide Reactivity Assay (DPRA) evaluates ability of chemicals to react with proteins. Profiler comprises 32 structural alerts derived from experimental cysteine depletion values. Classifies chemicals as low, moderate or highly reactive depending on percentage cysteine depletion
DPRA lysine peptide depletion	As above but based on lysine depletion – comprises 24 structural alerts. Classifies chemicals as low, moderate or highly reactive or non-reactive
Estrogen receptor binding	The profiler is based on structural and parametric rules from the literature. Classifies chemicals as non-binders or binders (weak/moderate/strong) depending on molecular weight and structural characteristics; includes 11 categories
Hydrolysis half-life (K_a, pH 7) Hydrowin	Profiler based on HYDROWIN; estimates half-life based on total acid-catalysed hydrolysis rate constant at pH 7
Hydrolysis half-life (K_a, pH 8) Hydrowin	Profiler based on HYDROWIN; estimates half-life based on total acid-catalysed hydrolysis rate constant at pH 8
Hydrolysis half-life (K_b, pH 7) Hydrowin	Profiler based on HYDROWIN; estimates half-life based on total base-catalysed hydrolysis rate constant at pH 7
Hydrolysis half-life (K_b, pH 8) Hydrowin	Profiler based on HYDROWIN; estimates half-life based on total base-catalysed hydrolysis rate constant at pH 8
Hydrolysis half-life (pH 6.5–7.4)	Estimates half-lives of organic chemicals under neutral or nearly neutral conditions, ambient or room temperature and atmospheric pressure. Classifies as very slow, slow, moderate, fast, very fast and extremely fast
Ionization at pH = 1	Calculates concentration of ionised species at pH = 1 using ionisation of strongest acidic and basic sites only. Requires pre-calculated pKa values therefore only applicable to compounds within the Toolbox database
Ionization at pH = 4	As above but for pH = 4
Ionization at pH = 7.4	As above but for pH = 7.4
Ionization at pH = 9	As above but for pH = 9
Protein binding by OASIS v1.1	Developed as part of the TIMES model for skin sensitisation. Identifies structural alerts responsible for eliciting effects as a result of protein binding, such as skin sensitisation; includes 85 categories. The mechanisms are developed using existing knowledge of reaction mechanisms

Table 4.2 (*Continued*)

Profiler	Further information
Protein binding by OECD	Direct acting structural alerts from literature compiled within a mechanistic chemistry framework. Enables category formation at the structural or mechanistic alert level. Mechanistic alerts, derived from identifying common reactivity sites on chemicals with a structural alert. Includes 102 categories, 16 mechanistic alerts relating to 52 structural alerts.
Protein binding potency	Developed using reactivity data measuring covalent binding with thiol group of glutathione. Includes 49 categories relating to the Michael acceptor mechanism and 46 categories relating to the S_N2 mechanism. Classifies chemicals as extremely reactive, highly reactive, moderately reactive, slightly reactive, suspect and non-reactive
Superfragments	Based on an algorithm developed by BioByte whereby isolating carbons *i.e.* those which act as a high-threshold barrier to the movement of electrons are identified and the remaining fragments are denoted as polar or "simple fragments". A superfragment consists of a combination of simple fragments that are in such close proximity that their solvation behaviour is affected. It can be defined as the "largest electronically-connected substructure"
Toxic hazard classification by Cramer (original)	The Cramer classification scheme uses a decision tree approach to make an estimation of a Threshold of Toxicological Concern (TTC) using chemical structures and estimated human intake. 33 structural rules are used to classify the substances into one of three classes: Class I "low toxicity" simple structures with known metabolism. Class II "intermediate toxicity" structures that are less innocuous than those in Class I do not possess features suggestive of high toxicity. Class III structures that do not indicate safety and may possess high toxicity
Toxic hazard classification by Cramer (with extension)	Modified version of above
Ultimate Biodeg	Classifies chemicals into persistence categories of (0–1 day, 1–10 days, 10–100 days and > 100 days) based on a conversion of Biological Oxygen Demand data to half-life data
Endpoint Specific Profilers	
Acute aquatic toxicity classification by Verhaar	Uses structural information to classify organic compounds into one of four classes for acute aquatic toxicity: inert (baseline toxicity); less inert; reactive; and specifically-acting chemicals
Acute aquatic toxicity MOA by OASIS	Classifies chemicals according to acute aquatic toxicity mode of action using structural information

Table 4.2 (*Continued*)

Profiler	Further information
Acute aquatic toxicity classification by ECOSAR	ECOSAR was developed by the US EPA and contains structure-activity relationships (SARs) used to predict the aquatic toxicity. The profiler in the Toolbox mimics the structural definition of the classes within ECOSAR enabling chemical classification but does not estimate toxicity
Bioaccumulation metabolism alerts	Estimates whole body primary biotransformation half-lives and biotransformation rate constants for organic chemicals in fish
Bioaccumulation metabolism half-lives	Uses estimates of biotransformation rates to classify chemicals with very slow, slow, moderate, fast, or very fast biotransformation rates
Biodegradation fragments (BIOWIN MITI)	Uses structural alerts used by the MITI Biodegradation Probability Models to estimate probability of rapid aerobic and anaerobic biodegradation in the presence of environmental microorganisms
Carcinogenicity (genotox and non genotox) alerts by ISS	Uses a decision tree to estimate carcinogenicity based on 55 structural alerts (35 mainly genotoxic carcinogenicity alerts that were previously encoded within the Toxtree software and 20 new alerts predominantly non-genotoxic). Alerts are based on molecular functional groups or substructures; includes 58 categories
DNA alerts for AMES, MN and CA by OASIS v1.1	Based on AMES mutagenicity models in TIMES system; identifies requirements for mutagenicity incorporating molecular flexibility and metabolic activation; includes 49 categories
Eye irritation/corrosion exclusion rules by BfR	A rulebase for eye irritation and corrosion developed by the German Federal Institute for Risk Assessment (BfR) and collaborators that uses structural alerts and physical-chemical exclusion criteria (lipid solubility, octanol:water partition coefficient, aqueous solubility, melting point and molecular weight) to determine which chemicals do not show eye irritation or corrosion potential; includes 31 categories.
Eye irritation/corrosion inclusion rules by BfR	17 structural alerts (categories) are defined giving inclusion rules for chemicals with potential for eye irritation and corrosion
In vitro mutagenicity (Ames test) alerts by ISS	A decision tree approach based on the Toxtree software using 30 structural alerts for mutagenicity based on substructure or molecular functional groups
In vitro mutagenicity (micronucleus) alerts by ISS	Based on the ToxMic rulebase in Toxtree that uses 35 structural alerts associated with induction effects in the micronucleus assay. Incorporates knowledge based on mechanisms of toxicity and chemical structure
Keratinocyte gene expression	22 categories developed based on the KeratinoSens assay which identifies chemicals inducing expression of luciferase reporter gene; detects electrophilic chemicals giving classifications of very high, high, moderate and low gene expression

Table 4.2 (*Continued*)

Profiler	Further information
Oncologic primary classification	Uses molecular definitions to mimic structural criteria of potential carcinogens, as indicated by US EPA's Oncologic software. Identifies 48 categories but does not predict carcinogenicity
Protein binding alerts for skin sensitisation by OASIS v1.1	Developed as part of the TIMES model for skin sensitisation. Identifies structural alerts responsible for eliciting effects as a result of protein binding, such as skin sensitisation; includes 85 categories
rtER expert system ver.1 USEPA	Identifies potential to bind to rainbow trout Estrogen Receptor
Skin irritation/corrosion exclusion rules by BfR	A rulebase for skin irritation and corrosion developed by the German Federal Institute for Risk Assessment (BfR) and collaborators that uses physical-chemical exclusion criteria (lipid solubility, surface tension, octanol:water partition coefficient, vapour pressure, aqueous solubility, melting point and molecular weight) to determine which chemicals do not show skin irritation or corrosion potential; includes 35 categories
Skin irritation/corrosion inclusion rules by BfR	40 structural alerts providing inclusion rules for classifying chemicals likely to cause irritation and/or corrosion of the skin
Empiric profilers	
Chemical elements	Organises chemical elements in the periodic table into 18 groups; includes 34 categories
Groups of elements	Organises chemical elements in the periodic table into 9 categories
Lipinski rule OASIS	Application of Lipinski "rule of five" developed to ascertain the likelihood of a chemical possessing good oral absorption properties (based on knowledge of hydrogen bonding, molecular weight and octanol:water partition coefficient
Organic functional groups[a]	Specific groups of atoms or bonds that define the chemistry of a compound are identified. The profiler can be used to identify structurally similar chemicals; includes 484 categories
Organic functional groups (US EPA)	This profiler is for identifying functional groups derived from the fragment library of the US EPA's KOWWIN program within EPISUITE; includes 467 categories
Organic functional groups (nested)	Profiler based on the organic functional groups[a] profiler above, however (nested) does not display functional groups that are part of larger functional groups
Organic functional groups Norbert Haider (Checkmol)	Identifies the presence of 204 functional groups recognised by the Checkmol program
Tautomers unstable	Developed on available data and theoretical calculations for tautomer forms in water and gas phase. Unstable tautomeric forms presented as individual categories; includes 158 categories

Table 4.2 (*Continued*)

Profiler	Further information
Toxicological Profile Repeated dose (HESS)	Gives category boundaries expected to induce similar toxicity for oral repeat dose based on data in Hazard Evaluation Support System (HESS); includes 33 categories

endpoint data are employed in the read-across prediction. However, it is essential to first establish which chemicals should be included in the final category. Again, this process is subjective and requires expertise to make the selection of chemicals.

Step (4) is the category definition step where chemicals are grouped together according to given criteria. This can be performed in an iterative manner until the user is confident that the category contains only true analogues of the query chemical. Chemicals are grouped based on structural and/or mechanistic similarity. If the mechanism of action is known then chemicals should be grouped based on descriptors or structures related to that mechanism; chemicals identified as being structurally dissimilar (either by visual inspection by the user or by subsequent refinement of the category using additional sub-categorisation criteria within the software) can later be excluded from the category. If the mechanism of action is unknown then chemicals can be grouped according to their structural features. Using profilers based on the presence of specific organic functional groups ensures a defined structural domain for the category. Once a category has been defined and populated with sufficient structures and their associated (toxicological) data a read-across prediction of activity for the unknown chemical can be made. Case studies 1–4 in Chapter 6 provide examples of the iterative process of sub-categorisation, using the Toolbox, to ensure a well-defined, robust category is obtained.

Step (5) is to fill the data gap *i.e.* to provide an *in silico* prediction of activity where no data are available for the target chemical. Prediction can be performed using trend analysis (useful for quantitative predictions where a large number of analogues are available), quantitative structure-activity relationship (QSAR) models (where no suitable analogues are available) or by read-across. For a read-across prediction, data that have been gathered on analogues in the category are used to make a prediction of the activity of the target chemical. Read-across is useful where quantitative predictions are required that can be based on a small number of analogues in the category or where a qualitative (or semi-quantitative) prediction is required (active, inactive, weakly active *etc*). Case studies 1–4 in Chapter 6 provide examples of categories formed and the read-across predictions made using the data available for other chemicals in the category.

Step (6), the final stage in the workflow of the Toolbox, relates to reporting of the prediction. For an *in silico* prediction to be acceptable, particularly in a

Table 4.3 Metabolic profilers available within the OECD QSAR Toolbox version 3.1.

Documented Metabolism	Further Information
Observed mammalian metabolism	Metabolic information for 100 chemicals (630 studies) metabolised in different mammals
Observed microbial metabolism	Information from 551 chemicals degraded by microorganisms
Observed rat *in vivo* metabolism	Metabolic information for 647 chemicals metabolised in rodents (predominantly rats)
Observed rat liver S9 metabolism	Metabolic pathway for 261 chemicals using rodent (predominantly rat) liver microsomes and S9 fractions
Simulated Metabolism	
Autoxidation simulator	Uses an autooxidation (AU) model for spontaneous free radical oxidation of chemicals at nearly neutral (pH 7–7.5) or slightly alkaline pH (8–9)
Autoxidation simulator (alkaline medium)	As above but pH 10.2–11.5
Dissociation simulation	Module developed within Laboratory of Mathematical Computing, Bourgas
Hydrolysis (acidic) simulator	Predicts hydrolysis products of specific organic chemicals at acidic pH, ambient or room temperature and atmospheric pressure
Hydrolysis (basic) simulator	Predicts hydrolysis products of specific organic chemicals at basic pH, ambient or room temperature and atmospheric pressure
Hydrolysis (neutral) simulator	Predicts hydrolysis products of specific organic chemicals at neutral or nearly neutral pH, ambient or room temperature and atmospheric pressure
Microbial metabolism simulator	Implementation of the CATABOL simulator for microbial metabolism using abiotic and enzyme-mediated reactions in a structured format
Rat liver S9 metabolism	Simulator (transformation table) for 509 structurally generalised biotransformation reactions characteristic of *in vitro* liver microsome and S9 fractions
Skin metabolism simulator	Estimates skin metabolism using a simplified mammalian liver metabolism simulator

regulatory setting, a clear and justifiable rationale for the prediction must be available. Within the Toolbox several options are available for recording all of the steps and processes that were undertaken in making the prediction. These include the QSAR Model Reporting Format (QMRF), the QSAR Toolbox category report and the QSAR Toolbox prediction report. The reports are customisable based on the user's requirements and are an essential component

Table 4.4 Toxicological categories and their associated effects following repeat dose testing as identified in the HESS category library.[4]

Toxicological Effect	Associated Categories
Haemolytic anaemia	Azobenzenes; diphenyl disulphides; hydrazines; oximes; nitrobenzenes; anilines; N-Alkyl-N'-phenyl-p-phenylenediamine; ethyleneglycol alkylethers; o/p-aminophenols
Thyrotoxicity	Imidazole-2-thione derivatives
Neurotoxicity	Acrylamides; organophosphates
Hepatotoxicity	Aliphatic nitriles; hydroquinones; aromatic hydrocarbons; nitrobenzenes; anilines; halobenzenes; p-alkylphenols; halogenated aliphatic compounds
Renal toxicity	p-Aminophenols; halobenzenes
Lipidosis of adrenocortical cells	Phenyl phosphates
Hepatobiliary toxicity	4,4'-Methylenedianilines/benzidines
Alpha 2u-globulin nephropathy	Aliphatic/alicyclic hydrocarbons
Mucous membrane irritation	Aliphatic amines; phenols
Less susceptible	Benzene or napthalene sulphonic acid
Testicular toxicity	Ethylene glycol alkylethers; nitrobenzenes; phthalate esters
Toxicity to urinary system	Benzene sulphonamide
Mitochondrial dysfunction	Nitrophenols/halophenols

of the documentation, as they may be required for submissions to regulatory agencies (see Chapter 7).

The OECD QSAR Toolbox has been designed specifically for the purpose of grouping chemicals into categories and making read-across predictions, however there are other useful tools available for grouping chemicals or providing information on chemical similarity and these are described below.

4.4 The Hazard Evaluation Support System (HESS)

Predicting repeat dose toxicity is one of the most challenging areas where *in silico* tools are applied due to the complexity of processes involved in the whole body system. There has been little success in applying traditional SAR or QSAR methods to these endpoints, however, predictions using category-based approaches that relate to mechanistic information are a more promising approach. Sakuratani *et al.* discuss the use of "toxicological categories" within HESS as a method to obtain read-across predictions for repeat dose toxicity.[4] HESS is a bespoke tool for evaluating chemicals for repeat dose toxicity using the category approach. HESS is freely available via the website of the National Institute of Technology and Evaluation (NITE), Japan, (http://www.safe.nite.go.jp/english/kasinn/qsar/hess-e.html) and is incorporated into the OECD QSAR Toolbox (see Section 4.3).

Development of the software involved analysis of repeat dose toxicity data for 500 chemicals administered to male and female rats over a dosing period of 28–120 days. Data from the literature were used to identify classes of chemicals that elicited similar toxicity effects. Boundaries for the categories were ascertained by investigation of the chemical space associated with the category members. The mechanism underlying the toxicity was described in a simplified Adverse Outcome Pathway (AOP). AOPs, as discussed in Chapter 3, provide a framework to link a molecular initiating event with a given biological response. Analysis of the data led to 33 toxicological categories being identified associated with 14 types of toxicity, summarised in Table 4.4.

HESS uses a similar workflow to the Toolbox (outlined above). It is compatible with the Toolbox, which contains a profiler of 33 categories based on the HESS tool which relates only to repeat dose toxicity data. HESS also incorporates a database of raw data for repeat dose toxicity and metabolic information.

Case study 5 in Chapter 6 details the use of the HESS tool for making a read-across prediction.

4.5 Toxmatch

Read-across is based around the grouping of similar molecules. Molecules cannot be classified as similar in an absolute sense, but they may be considered "similar" with respect to a given property. There are many possible methods to determine the similarity of molecules and form groups based on that particular measure of similarity. The Toxmatch program was commissioned by the European Commission's Joint Research Centre (JRC) and developed by Ideaconsult, Sofia. Version 1.07 was released in 2009 and can be freely downloaded, along with the User Manual (from which much of the information below was obtained) and other supporting information, from the JRC website (http://ihcp.jrc.ec.europa.eu/our_labs/computational_toxicology/qsar_tools/toxmatch).[5] Toxmatch is a flexible, open-source software program that can be used to group chemicals together and predict activity or classify new chemicals into an appropriate group based on similarity measures.

Training set data can be uploaded into Toxmatch in a range of formats including CML, CSV, HIN, ICHI, INCHI, MDL MOL, MDL SDF, MOL2, PDB, SMI, TXT and XYZ file types. Toxmatch can be used to generate a small number of descriptors but a user can also upload descriptors, obtained using alternative software, into the Toxmatch environment. The software uses knowledge of the endpoint data (*i.e.* a supervised training technique) to classify new chemicals and predict activity. There are six datasets for toxicity preloaded into Toxmatch, these are: the DSSTox EPA Fathead Minnow Acute Toxicity dataset; a bioconcentration factor (BCF) dataset; a skin sensitisation dataset obtained using the mouse local lymph node (LLNA) assay; a skin irritation and corrosion dataset; the Instituto Superiore di Sanita chemical carcinogens (ISSCAN) database, which includes both rat and mouse

Table 4.5 Similarity measures available in Toxmatch (adapted from Toxmatch User Manual).[5]

Similarity Measure*	Description
Euclidean distance (descriptors, kNN)	Average Euclidean distance between selected descriptors for the query chemical and k most similar chemicals from the dataset. More similar chemicals have lower Euclidean distance
Hodgkin–Richards index (descriptors, kNN)	Average Hodgkin–Richards index between selected descriptors for the query chemical and k most similar chemicals from the dataset. More similar chemicals have higher Hodgkin–Richards index
Cosine similarity index (descriptors, kNN)	Average Cosine index between selected descriptors for the query chemical and k most similar chemicals from the dataset. More similar chemicals have higher Cosine index
Tanimoto distance (descriptors, kNN)	Average Tanimoto distance between selected descriptors for the query chemical and k most similar chemicals from the dataset. More similar chemicals have higher Tanimoto index
Tanimoto distance (Fingerprints, kNN)	1024 bit length hashed fingerprints can be generated based on Daylight's fingerprint theory. Calculates average Tanimoto index between 1024 bit fingerprints of query chemical and k most similar chemicals from dataset consensus fingerprint. More similar chemicals have higher Tanimoto index
Hellinger distance (atom environments, kNN)	Atom environments are regarded as fragments surrounding each atom in a molecule up to a set level. Calculates average Hellinger distance between atom environments of query chemical and k most similar chemicals. More similar chemicals have a higher index

*Note the nearest neighbours identified by each similarity measure will be different.

carcinogenicity studies in addition to Ames test mutagenicity data. The pre-loaded training sets have previously been categorised *e.g.* by mode of action as defined in the Verhaar scheme. Alternatively, users can input their own training set with relevant endpoint data from which new groups can be built. Query chemicals can then be classified into an appropriate group for toxicity prediction.

Toxmatch can be used to quantify the level of similarity between two chemicals based on similarity in descriptor space (Euclidean distance, Hodgkin-Richards index, Tanimoto index or cosine-like (Carbo) index) or structural similarity (Tanimoto index, Hellinger distance, or Maximum Common Substructure Similarity (MCSS)). Once the similarity index between two compounds is established, a similarity index can be determined between an individual chemical and a set of chemicals. This can be performed between a representative chemical within the dataset and the query chemical. This approach uses the Tanimoto distance (Fingerprints, kNN) method or the Hellinger distance (atom environments, summary atom environment method) method. Note that Chapter 2 discusses use of Tanimoto coefficients as a means

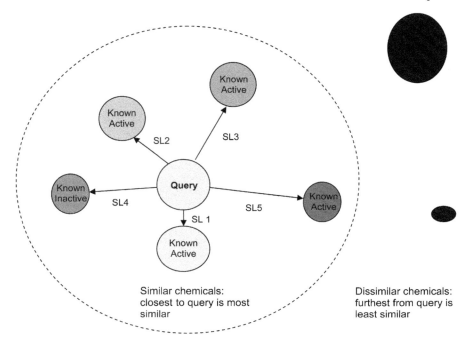

Figure 4.2 Measure of similarity between query chemical and its nearest neighbours (adapted from Toxmatch User Manual).[5] SL is the "Similarity Level" between the query chemical and its neighbours.

to assess similarity as an example of chemometrics-based category formation. Another approach is to determine the average similarity between the query chemical and its k nearest neighbours (by default 10 nearest neighbours are selected, but the user can specify an alternative number). Table 4.5 provides a summary of the options for determining similarity between one chemical and a set of chemicals within Toxmatch.

Toxmatch can use this similarity information to predict (read-across) the activity of a query chemical, based on activity of other chemicals in the category. The category comprises a number of nearest neighbours (as defined by the user) and activity is based on a weighted average of the activity of these nearest neighbours (where the most similar chemical has the highest weighting). Alternatively Toxmatch can be used as a tool to group chemicals together based on their calculated degree of similarity. A query chemical can then be assigned to one of the groups based on its similarity to other members of the group *i.e.* it will be assigned to the group that contains most of its k nearest neighbours. Figure 4.2 shows a diagrammatic representation of a Toxmatch prediction for activity based on similarity measure and k nearest neighbours.

Calculating the similarity of a query chemical to a training set is used to predict activity or classify the query chemical, as outlined above. Calculating similarity to a test set can be used to compare two datasets. Where possible similarity should be assessed based on criteria relevant to the endpoint. Where mechanistic knowledge is available this should be used to perform grouping; where this is not available similarity based on relevant descriptors or structural features can be used. Users can establish their own "cut-off" criteria *i.e.* above what level of similarity between two (or more compounds) should they be classed as truly similar?

Case study 6 in Chapter 6 describes the use of Toxmatch to form structural categories from which predictions of toxicity (teratogenicity) were made.

Sections 4.3–4.5 above have described tools available to support category formation and read-across that have been exemplified using the case studies in this book. Subsequent sections in this chapter briefly describe other tools that may assist with placing chemicals into rationally defined groups or identifying "similar" molecules.

4.6 Toxtree

The Toxtree software was commissioned by the JRC and developed by Ideaconsult, Sofia. Version 2.5.1 is available via the Sourceforge open source software site (http://sourceforge.net/projects/toxtree/files/) from here it can be freely downloaded along with supporting information and a user manual.[6] Toxtree was designed to estimate toxic hazard using a decision tree approach. As indicated in Table 4.2, much of the knowledge from the decision tree approach of Toxtree has been incorporated into existing profilers within the Toolbox. The decision trees categorise chemicals based on specific (physico-) chemical features and/or structural alerts. In this way chemicals within a dataset of interest can be classified into different groups. Toxtree is currently distributed with 14 plug-ins representing different toxicity endpoints: Cramer rules (with extensions) for Threshold of Toxicological Concern; Verhaar scheme (and modified Verhaar scheme) for aquatic mode of toxic action; skin irritation prediction; eye irritation prediction; Benigni/Bossa rules for mutagenicity and carcinogenicity; STructural Alerts for Reactivity in Toxtree (START) biodegradation and persistence; structural alerts for identification of Michael acceptors; structural alerts for skin sensitisation; Kroes Threshold of Toxicological Concern Decision Tree; SMARTCYP (cytochrome P450- mediated drug metabolism prediction); structural alerts for the *in vivo* micronucleus assay in rodents; and structural alerts for functional group identification. Users may modify the decision trees using their own structural rules if required or devise new decision trees. Input of structures in single or batch mode is similar to that for Toxmatch. Results for the classification of the compound can be displayed in simple or verbose format. It is possible to view the complete tree via which the decision for a classification has been derived and a verbose explanation as to why the chemical has been placed into a particular class. A file containing a dataset of compounds can be split into subsets based on the results of a chosen decision tree, in this way compounds within a dataset can be separated into defined groups.

4.7 AMBIT

AMBIT was developed with funding from industry via a European Chemical Industry Council Long Range Initiative (CEFIC-LRI), and is freely downloadable via the Sourceforge website (http://ambit.sourceforge.net/). It comprises a relational database of compounds (including over 450,000 chemical structures and their identifiers), associated properties, QSAR models, references and tables of data including pre-calculated fingerprints (that allow substructure and similarity searching to be performed more rapidly). Similarity searching of molecules based on Tanimoto coefficient values (see Chapter 2) can be performed and substructure searches for similar molecules are also possible. Jeliazkova *et al.* (2010) describe the use of a workflow utilising AMBIT to identify analogues within a dataset. This is summarised as follows:[7]

(i) The starting set of structures are defined using CAS/EINECS number, names, SMILES, MOL or SDF files or using a structure editor
(ii) An analogue search is performed (by default hashed fingerprints are compared by Tanimoto distance)
(iii) The results are displayed and the user can decide to restrict further queries within the set of selected structures
(iv) Substructure search performed by user-defined fragments
(v) Results can be further filtered by compound profiles (*e.g.* by experimental or calculated data *e.g.* octanol:water partition coefficient)
(vi) Selected structures can be grouped into typical chemical classes or clustered to identify groups of analogues.

4.8 Leadscope

Databases and data-mining tools are available from several software vendors. One example is the toxicity database available, commercially, from Leadscope (http://www.leadscope.com). This comprises 180,000 records with associated data from over 400,000 toxicity studies. The types of toxicity data that may be available for chemicals include; chronic and acute data; carcinogenicity, reproductive and developmental toxicity, genetic toxicity, irritation *etc.* Searches based on structures, sub-structures or similarity measures can be used to identify potential analogues. There are also routines for creating subsets of data to work within a defined region of chemical space. Large data sets may be classified using mechanistically-driven structural classifiers and datasets mined for compounds possessing particular structural alerts. In this way the software can be applied for the purpose of identifying compounds sharing common features that may form the basis for grouping.

4.9 Vitic Nexus

This is a chemical database and information management system, released by the not-for-profit organisation Lhasa Limited, Leeds. The Vitic Nexus database holds data records for over 12,000 structures and is continually expanding. There are options to include searches of in-house data or link searches to external sources such as ToxNet or Google at the user's discretion. The database can be searched by entering structures via Marvin Sketch (which is integrated into the program) or entering a molfile, SD file, SMARTS, InChI or CAS registry number. Endpoint data can be obtained by entering a structure and searching for information on exact matches. Alternatively, and more appropriately for building a category, it is possible to search for information on substructures or for compounds that are "similar" to the input compound. Similarity can be set at a level between 0% and 100%. Having completed a search a results gallery shows all of the compounds meeting the initial criterion, from this the user can select which compounds to investigate further. The results explorer details the endpoint data available for the compounds selected. The advantage of the Vitic Nexus database is the detailed, high quality, information that is available for each compound. Endpoint data include experimental results reported within a toxicological ontology that makes it possible to select a specific outcome *e.g.* clinical chemistry changes, general toxicity, gross necroscopic findings, dose–response information *etc*. This may be supplemented by experimental protocols, literature references and supporting information from the studies. Chemicals of interest can be grouped according to given criteria, such as those containing a given substructure or similarity to a target. Detailed endpoint information, where available, can be used to aid read-across predictions for toxicity.

4.10 ChemSpider

ChemSpider (http://www.chemspider.com), owned by the Royal Society of Chemistry, is a freely available chemical database that currently holds information on over 28 million compounds and their properties, collated from over 400 data sources. Links to the original data sources are provided enabling checking of data where necessary. More compounds and data are continually added to this database and its use as a data resource is discussed further in Chapter 5. The database can be searched by entering a chemical structure or an identifier (systematic name, trivial name, InChI, SMILES string or registry number). Searches can be performed to find compounds exhibiting common properties such as the presence of specific elements within a chemical, calculated properties (*e.g.* searches performed to identify compounds within a given log P range) or Lasso similarity (ligand activity in surface similarity order) or finding compounds that are associated with a particular data source.

Whilst it is possible to search for information on a given chemical, it is also possible to search for chemicals related to a given target and so identify potential members of a group. This can be performed by using the substructure or similarity search features. It is possible to search for exact structures, substructure or by similarity measures (*e.g.* Tanimoto, Tversky, Euclidean) at a level of greater than or equal to 50%, 70%, 80%, 90%, 95% or 99% similarity. Once similar compounds have been selected, the database can then be investigated for the information held (endpoint data) on those chemicals. This may help identify a potential group of structurally similar chemicals for which toxicity data may be available.

4.11 ChemIDPlus (Advanced)

This is another freely available web-based resource (available at http://chem.sis.nlm.nih.gov/chemidplus) containing records for over 390,000 chemicals cited in the National Library of Medicine of the Unites States. Structures are available for more than 300,000 chemicals and searches can be performed by entering names, synonyms, CAS registry number, molecular formula, structure, toxicity, physical properties *etc*. This resource enables searches to be performed to identify similar structures to a given target molecule at specified values for similarity (*e.g.* 50%, 60%, 70%, 80% or 90% similarity). It is also possible to search for given substructures and compounds present as salts, hydrates, mixtures, *etc.* as required. Available toxicity data for compounds of interest can be retrieved and searches refined based on organism, route of administration, toxicity elicited or physical properties; supporting references are also provided.

4.12 Analog Identification Methodology (AIM)

The AIM software was developed by the US Environmental Protection Agency as part of the Sustainable Futures Initiative and is freely downloadable from the EPA website (http://www.epa.gov/opptintr/sf/tools/aim.htm). The tool was designed such that analogues of a chemical of interest could be identified and the software would indicate the publicly available sources from where toxicity data for the analogues could be obtained (the data are not contained within the software itself). Structure searching is performed using CAS registry numbers, SMILES strings or (sub)structure searching. Structural analysis is performed using 700 pre-defined atoms, groups and superfragments and comparing these to a database of 86,000 chemicals for which toxicity data are publicly available. The toxicity data for these analogues can then be used for read-across. At time of writing the US EPA is collating comments on the utility of this tool that may be considered for future updates of the software.

4.13 Use of Computational Workflows in Read-Across

A flow diagram showing the overall process involved in building a category and making a read-across prediction is given in Chapter 1, Figure 1.3. Whilst the current chapter has covered some bespoke tools for making read-across predictions following this process (*e.g.* OECD QSAR Toolbox) there are other tools or combinations of resources that also may be useful as they offer a more flexible, less prescriptive approach. An example is the development of computational workflows; these allow information and processes from diverse sources (*e.g.* different databases and software packages) to be fully integrated into a single predictive tool. The design and application of the workflow is entirely under the control of the developer or user. These workflows can be shared amongst users to ensure consistency and to enable further development of additional features. The user can decide which features to include or exclude in their own data analysis. Pipeline Pilot from Accelrys (http://accelrys.com/products/pipeline-pilot/) is a commercially available tool that enables such workflows to be built and shared. Another particularly useful tool is the open-source KNIME software (available from http://www.knime.org) which provides flexible graphical workflows for automating data analysis including accessing, transforming, processing, analysing and visualising data. Over 1,000 nodes are currently available; a network of users continually develop additional nodes. For the purposes of read-across an example of a workflow using KNIME is shown in Figure 4.3. This figure is based upon the workflow presented in Figure 1.3, but demonstrates how a KNIME workflow could be devised to assist in the process.

The benefit of such workflows is that they can not only ensure consistency between users but also permit flexibility in developing workflows adapted to particular needs. For example, users can incorporate their own profilers (*e.g.* profilers based on chemotypes associated with a given toxicity) or can incorporate databases of interest (such as in-house databases). The EU FP7 COSMOS project (http://www.cosmostox.eu/) is currently developing KNIME workflows for read-across for the prediction of toxicity.[8]

4.14 Conclusions

Category formation and read-across are increasingly being recognised as methods to provide reliable and justifiable predictions of toxicity that are important to users within industry and by regulators. Several useful tools have already been developed to assist in this process. This chapter has provided an overview of some of the tools currently available, including those specifically designed for this purpose, as well as other software and applications that could be usefully employed. Of the tools described herein, the Toolbox is the most directly applicable to making and documenting read-across predictions. However, the Toolbox requires a degree of expertise to be used successfully. In certain circumstances a less formalised and prescriptive method to find and

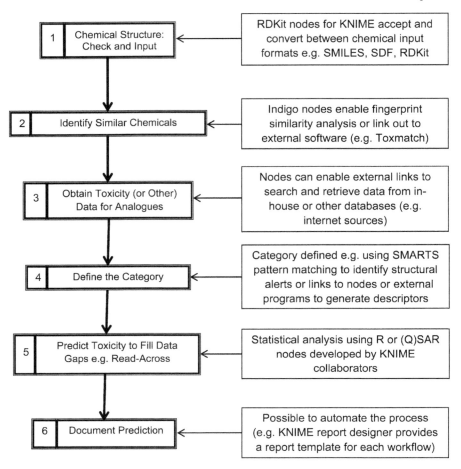

Figure 4.3 Developing a KNIME workflow for read-across including examples of the freely available nodes that could be applied. The boxes on the left side have been taken from Figure 1.3.

identify groups of chemicals may be useful *e.g.* using tools such as Toxtree, Toxmatch or AMBIT. Internet applications, such as ChemSpider (and to a lesser extent ChemIDPlus Advanced), allow the vast resources available via the internet to be investigated. For example substructure or similarity searching is now possible amongst millions of known chemicals. The use of computational workflows has also gained momentum recently and large numbers of nodes for these workflows are now being developed and shared amongst practitioners. As this area of science continues to develop, it is anticipated that a greater number of tools will become readily accessible to users.

Acknowledgements

Funding from the European Community's Seventh Framework Program (FP7/2007–2013) COSMOS Project under grant agreement no. 266835 and from Cosmetics Europe is gratefully acknowledged. Funding from the European Chemicals Agency (Service Contract No. ECHA/2008/20/ECA.203) is also gratefully acknowledged.

References

1. OECD QSAR Toolbox, *User Manual Getting Started*, Available from http://www.oecd.org/dataoecd/58/56/46210452.pdf (Accessed April 2013).
2. S. J. Enoch, C. M. Ellison, T. W. Schultz, and M. T. D. Cronin, A review of the electrophilic reaction chemistry involved in covalent protein binding relevant to toxicity. *Crit. Rev. Toxicol.*, 2011, **41**, 783.
3. S. J. Enoch and M. T. D. Cronin, Development of new structural alerts suitable for chemical category formation for assigning covalent and non-covalent mechanisms relevant to DNA binding, *Mutat. Res.–Genetic Toxicol. Environ. Mutagen*, 2012, **743**, 10.
4. Y. Sakuratani, H. Q. Zhang, S. Nishikawa, K. Yamazaki, T. Yamada, J. Yamada, K. Gerova, G. Chankov, O. Mekenyan and M. Hayashi, Hazard Evaluation Support System (HESS) for predicting repeated dose toxicity using toxicological categories. *SAR QSAR Environ. Res*, 2013, **24**(5), 617.
5. N. Jeliazkova, G. Patlewicz, A. Gallegos-Saliner, *Toxmatch User Manual*, Available from http://ihcp.jrc.ec.europa.eu/our_labs/predictive_toxicology/qsar_tools/toxmatch (Accessed April 2013).
6. Ideaconsult, *Toxtree User Manual*, Available from http://ihcp.jrc.ec.europa.eu/our-labs/predictive-toxicology/qsar-tools/toxtree (Accessed June 2013).
7. N. Jeliazkova, J. Jaworska and A. P. Worth, Open-source tools for read-across and category formation, in *In Silico Toxicology: Principles and Applications*, ed. M. T. D. Cronin and J. C. Madden, The Royal Society of Chemistry, Cambridge, 2010, p. 408.
8. M. D. Nelms, S. J. Enoch, E. Fioravanzo, J. C. Madden, T. Meinl, A.-N. Richarz, C. H. Schwab, A. P. Worth, C. Yang, M. T. D. Cronin, Strategies to form chemical categories from Adverse Outcome Pathways in *Towards the Replacement of in Vivo Repeated Dose Systemic Toxicity Testing. Volume 2*, ed. T. Gocht and M. Schwarz, Coach Consortium, Paris, France, 2012, p 263.

CHAPTER 5

Sources of Chemical Information, Toxicity Data and Assessment of Their Quality

J. C. MADDEN

School of Pharmacy and Chemistry, Liverpool John Moores University, Byrom Street, Liverpool, L3 3AF, England
E-mail: j.madden@ljmu.ac.uk

5.1 Introduction

Predicting human health effects or environmental toxicity of a compound inherently relies upon the availability of high quality data relating to the compound itself or to "similar" compounds, from which an inference can be made. The data required will depend upon the endpoint of interest and the nature of the query. Fundamentally, there are different types of data relating to compounds that may be useful: chemical data that relate to the identity and representation of the compound; descriptor data that relate to physico-chemical, structural or other properties of the compound; and data relating to its biological activity. For a given biological endpoint specific, quantitative toxicological data may be needed (for example tumour promoting activity relating to carcinogenicity) or qualitative indicators of activity or non-activity in a given assay may be appropriate. In other cases *in vitro* or *in chemico* data that are indicative of toxic potential may also be useful. These data can be used to make a read-across prediction of activity from similar compounds (in a given category) with known activity to those for which the endpoint data are unknown. However, grouping a collection of molecules together to build such a category requires information on what makes the chemicals "similar". This may use information on chemical structure, properties, similarity measures, fingerprints, metabolic profile, chemical reactivity *etc.* In reality, any data that allow chemicals to be grouped together in a rational manner can be useful for category formation. Within the chemical universe there is an ever-expanding knowledge base of properties of chemicals that can be explored and utilised. Data available include physico-chemical properties, reactivity measures, information regarding mechanism of action, *in vitro* and *in vivo* activity and many more. This chapter will cover aspects of obtaining relevant data for compounds of interest and how to assess the quality of those data in terms of accuracy, reliability and fitness-for-purpose. Hence, the emphasis of this chapter will be on where to find appropriate data and methods to assess data quality. Selecting which compounds should be the subject of the search (*i.e.* which compounds are considered as "similar" or belonging to a particular category) relates to the approach by which the category is to be formed and is the subject of Chapters 2 and 6.

5.2 Data Useful for Category Formation and Read-Across

There is a wide range of data available that could potentially be used for toxicity prediction using read-across. For predicting an adverse human health effect of a compound, human *in vivo* toxicity data for other compounds within the category would provide the most reliable information from which to make a prediction. However, as apical endpoint data in humans are unavailable in most instances, other data on the compounds within the category may also be useful. The list below is intended to be indicative only of the range of data that may be available and should not be considered to be an exhaustive list.

Potentially useful data include:

- apical *in vivo* data measured in other species (note route of administration and known inter-species differences need to be taken into account);
- organ level toxicity; histopathological findings from acute and repeated dose tests;
- absorption, distribution, metabolism and excretion (ADME) data including: permeability through biological membranes (*e.g.* Caco-2 permeability for oral absorption); skin flux measurements for dermal absorption; metabolic fate data including the nature of the metabolites formed and potential activity of the metabolites;
- *in vitro* activity (*e.g.* Ames tests for mutagenicity, microsomal stability assays, cytotoxicity tests *etc.*);
- information on biological mechanism (or mode) of action;
- information on mechanisms of chemical reactivity;
- reactivity assay data using biological surrogates (*e.g.* peptide or glutathione reactivity);

- binding studies for biological macromolecules (*e.g.* oestrogen or androgen receptor binding, DNA binding);
- environmental fate (including affinity for specific environmental compartments);
- structural similarity (analogues, presence of specific functional groups, size, shape, volume, molecular fingerprints);
- physico-chemical properties (*e.g.* logarithm of the octanol:water partition coefficient, aqueous solubility, volatility).

Using skin sensitisation as an example, data from the human repeated insult patch test may be most appropriate for category formation.[1] If these were unavailable, data from skin sensitisation tests in other species (*e.g.* the guinea-pig maximisation test or the mouse local lymph node assay) could also be used to make a prediction.[2,3] For this particular endpoint, much is understood regarding the underlying mechanisms of the process that has enabled specific *in vitro* assays to be developed. Three such assays were evaluated by Colipa (the cosmetics industry's trade association, now known as Cosmetics Europe) and have been submitted to ECVAM for prevalidation.[4] These are the Direct Peptide Reactivity Assay (DPRA), the Myeloid U937 Skin Sensitisation Test (MUSST) and the human Cell Line Activation Test (hCLAT); results from these assays may also be useful in modelling. The molecular initiating event in this process involves binding of a chemical to a skin protein via an electrophilic mechanism, hence chemical reactivity, measured using *in chemico* assays or estimated using quantum chemical calculations can also be useful for forming categories of compounds from which to make a prediction (such approaches for skin sensitisation are discussed in Chapters 2 and 6). Categories may be formed to help prioritise which chemicals should be selected for further toxicity tests if the available information is insufficient.

It is important to note that more than one type of data may be available for chemicals within a category *i.e.* different types of complementary data may be available and these may provide greater confidence in weight-of-evidence based predictions (see Section 7.3.1). The ToxCast program from the United States Environmental Protection Agency (US EPA) is an example of how a range of data can be combined to give an overall indication of toxic potential. The program was designed to investigate if multi-dimensional analysis of chemicals and their properties could be used to predict whole animal toxicity. Types of information include: physico-chemical properties; activity predicted using structure-activity relationships; biochemical properties from high throughput screening (HTS) assays; cell-based phenotypic assays; genomic analysis of cells *in vitro*; and responses in non-mammalian model organisms.[5] In the first "proof-of-concept" phase of the ToxCast program (completed in 2009) more than 300 chemicals (primarily pesticides, for which a large amount of traditional test data were available) were screened in over 235 HTS bioassays. Phase II of the program, in conjunction with the Tox21 project (http://epa.gov/ncct/Tox21/) is expanding the number of chemicals and assays used. This will provide a vast database of information which can be used to predict the toxic potential of chemicals and enable prioritisation for further testing. So-called "bioactivity signatures" or "toxicity signatures" elucidated using the combined results of such assays are a promising resource for information on chemicals within categories.

In addition, there are other measures of toxicity, that were not previously considered amenable to traditional *in silico* approaches, that are now being investigated for their potential use in category-based approaches. These include the No Observed (Adverse) Effect Level (NO(A)EL) and Lowest Observed Effect Level (LOEL). As these represent effects on the organism as a whole and are not specific to a given organ or mechanism, generally they cannot be modelled by statistical correlative approaches. However, there is potential to use such information within a category-based approach for toxicity prediction (see Section 6.2.5).

The information presented here indicates that there is a wide range of data that may be useful in developing categories or providing further support for a prediction of activity. There are many resources available from where such data may be obtained such as those described below.

5.3 Sources of Data

The first step in populating any type of category with data is determining which data exist. The amount and diversity of data available on chemical compounds is increasing on a daily basis. This section provides a useful starting point in searching for data but undoubtedly more resources will become available and up-to-date searches for new resources should be performed. Two key sources of information are in-house data and publicly available data.

5.3.1 In-house Data Sources

In-house data are very useful for developing *in silico* models, but particularly for developing datasets for read-across. This is because projects within a company often develop compounds for a specific purpose (possibly based on a lead compound) by investigation of a range of compounds that are classed as "similar". Hence it is possible to create a dataset from a pre-defined, relatively narrow and relevant region of chemical space. Other advantages are that the data are traceable, quality assurance can be readily checked, further information on experimental procedures (*e.g.* confirmation of dosing regimen, detailed protocols, vehicle or controls used, unusual observations etc) and other such queries can be taken directly to those who generated the original data. It may also be possible to request confirmatory assays to be performed within an iterative design/test cycle. This enables the reliability of the models, or methods used to generate the models, to be established increasing credibility and acceptability of predictions.

5.3.2 Public Data Sources

The scientific literature is an immense resource holding information on millions of compound. However, searching for specific data can be an extremely time-consuming process and in some cases it becomes necessary to develop a database to organise the information into a usable format. Data may be in the form of journal articles reporting results for compound(s) in a single study or series of studies. Results for multiple compounds from a single study may provide a complete data set where the information has been acquired on compounds investigated at the same time in the same laboratory, so reducing experimental variation in results. Journal articles are now frequently published with supplementary information being available from the publisher or authors. Subject to appropriate quality checks the supplementary information (*e.g.* spreadsheets of structures with associated data) can provide ready-made data sets for modelling. Compilations of results from diverse sources may be available in journal articles, websites or within databases.

Non confidential regulatory submissions can also provide relevant data. The European Chemicals Agency (ECHA) website (http://echa.europa.eu/) provides links to robust summaries of studies on chemicals. Similarly OECD Existing Chemicals Screening Information Data Sets (SIDS dossiers; available from http://www.inchem.org/pages/sids.html) are another source of information.

There are also resources for "cleaned" data sets which have already undergone certain quality control checks and therefore may be useful for modelling, for example the web pages of the Chemoinformatics and QSAR Society (http://www.qsar.org) provides links to a range of data sets, some providing structural information along with activity data. In terms of quality control, a "trust level" is indicated for the data sets. The website for the Inchemicotox project (http://www.inchemicotox.org) similarly provides toxicity data sets along with an assessment of their quality. (Assessment of data quality is discussed in more detail in Section 4.5). Cronin provides a list of well-established data sets and their sources.[6] These data sets can be searched to find chemicals that are "similar" to those in the category being formed.

Whilst toxicity data are clearly relevant for developing predictive models, other data from the literature can also be useful for building categories. Schwöbel *et al.* published a comprehensive review of chemical reactivity data obtained from literature spanning 80 years.[7] The database provides an openly available resource to assist in predicting reactive toxicity. It reports information on toxic effects, which are brought about by reactive mechanisms (particularly, those involving the formation of covalent bonds between electrophilic substrates and biological nucleophiles (*e.g.* nucleic acids and proteins). Such information is useful in forming chemically relevant categories, informed by potential mechanism of interaction within the body. Kalgutkar *et al.* provide a comprehensive review cataloguing all known bioactivation pathways of functional groups or structural motifs commonly used in drug design efforts.[8] Such information on metabolic pathways or biochemical reactions can also provide important insight into potential biological interactions for chemicals in a category.

Many international collaborative projects (former and on-going) have led to the compilation and curation of large data sets that also may be useful for populating categories. A major task within such projects is agreeing a strategy to collate, check and accurately record available information. Consequently, these are ever-expanding resources that can be usefully exploited by modellers saving duplication of effort; some of these resources are detailed below. The OECD QSAR Toolbox contains many databases donated by various collaborative partners, these provide a vast resource of useful data (details of which are given in Section 4.3).

Of the on-line freely available resources, ChemSpider and the Toxicity Reference Database (ToxRefDB) are particularly noteworthy. ChemSpider (http://www.chemspider.com) from the Royal Society of Chemistry, provides an online database with records for over 28 million structures (at time of writing), properties and associated information. This is continuously updated with more data as they become available. Ensuring the accuracy of these records is a major undertaking and this resource offers users the opportunity to comment on records (for example errors or updates can be reported). ChemSpider curators act upon such reports making corrections as necessary and recording action taken in a curation update. This ensures a traceable record of curated data and updates. ToxRefDB (http://www.epa.gov/ncct/toxrefdb/), maintained by the US Environmental Protection Agency (US EPA), is a publicly available collation of the results of thousands of *in vivo* animal studies (covering approximately 30 years and $2 billion worth of tests). Details of the study protocols are also available within ToxRefdb and searches can be linked to other resources such as ACToR (see below) and the ToxCast program. VITIC Nexus is another database (from the not-for-profit company Lhasa Ltd, Leeds) which is currently under development and includes detailed information on toxicity testing results, experimental protocols and supporting information that can be searched using structures, substructures or similarity. VITIC Nexus has the option to add in data from other sources, such as in-house databases, and searches can be linked to external data sources such as Google or the toxicology data network (ToxNet; developed by the US National Library of Medicine) to enable multiple databases to be searched simultaneously.

Other major resources are referred to as "global (meta)portals", such as AcTOR (http://actor.epa.gov/actor/faces/ACToRHome.jsp) from the US EPA. This houses all publicly available chemical toxicity data from over 1000 public sources. Similarly the Organisation for Economic Cooperation and Development (OECD) has developed the e-Chemportal (http://www.echemportal.org) providing publicly available information on physical and chemical properties, toxicity, ecotoxicity, environmental fate and behaviour, via links to a host of electronic resources. Commercial databases are another useful resource although there is a cost associated with these.

As the number of potential resources increases, technological advances are needed in order to exploit the data fully. Recent technological innovations have led to increased accessibility of data, for example ElementalDB is a chemical substructure and similarity based searching application incorporating the ChemBL 15 dataset and is now available as a Mobile Chemistry App for iPhone and iPad devices (http://www.dotmatics.com/); future developments will increase the number of databases searchable using such devices. Developments in data and text mining are also helping to solve some of the issues in finding relevant data from within the immense amount of information available. For example, Fourches *et al.* used automated text mining, with limited manual curation, to gather liver toxicity data, from MEDLINE extracts, for 951 compounds.[9] These data were subsequently used for developing predictive models for liver toxicity, demonstrating a promising approach for future data collation and modelling efforts.

The resources available are continually expanding, hence it is not possible to comprehensively review these within one chapter, however, Table 5.1 provides a summary of some potentially useful resources (note that this is not an exhaustive list).

5.4 Strategies for Data Collection

When collecting data to populate a category, a pragmatic approach with a clear rationale must be applied. Chemicals incorporated into the category (and their associated data) must fit the specific profile of the category. One approach is to begin searching for compounds using broad constraints to obtain a larger data set. This data set is then iteratively pruned to produce smaller, more selective sub-categories of true analogues (the process of sub-categorisation). The term "analogues" is defined here as those compounds that produce their effects by the same mechanism. Attempting read-across from a large category containing too many chemicals may lead to inaccurate predictions as other effects may come into play. For example a category built on compounds containing a specific functional group may need to be further restricted to compounds falling within a given molecular weight range. Larger molecules may not reach the target site or their activity may be hindered due to steric effects. Chapters 2 and 6 consider the approaches by which categories may be formed and factors that should be taken into consideration when building a category of the appropriate size. For grouping and read-across approaches a smaller, more closely related group of compounds, for which accurate data are available, is more useful than a larger and more diverse data set. Whatever the source of the data and however many compounds are used to populate the category, it is essential that the data are assessed for quality in terms of accuracy, reliability and fitness-for-purpose.

Table 5.1 Potential sources for toxicity, reactivity, chemical property and structural data.

Data Resource	Information	Reference
Literature information (reviews and summaries)		
Schwöbel et al.	Information on toxic effects mediated by mechanisms (emphasis on potential for covalent bond formation between electrophilic chemicals and biological nucleophiles)	Ref [7]
Kalgutkar et al.	Comprehensive review of all known bioactivation pathways of functional groups or structural motifs commonly used in drug design	Ref [8]
Enoch et al.	A review of the literature defining the structural alerts associated with covalent protein binding and the associated electrophilic reaction chemistry	Ref [10]
Enoch et al.	Structural alerts for chemical category formation for assigning covalent and non-covalent mechanisms relevant to DNA binding.	Ref [11]
OECD Screening Information Data sets (SIDS) dossiers	Comprehensive reviews for (eco)toxicity endpoints	www.inchem.org/pages/sids.html
ECHA summaries	Robust summaries of general information, classification and labelling, physical and chemical properties, toxicological and ecotoxicological data, environmental fate etc	http://echa.europa.eu/
Web resources for data sets		
The Chemoinformatics and QSAR Society	Provides links to QSAR data sets covering many endpoints and including structures for compounds	www.qsar.org/resource/datasets.htm
Projects involving data compilation		
ACuteTox	Compilation of *in vitro* and *in vivo* data, relating to acute toxicity, for a reference set of compounds	www.acutetox.org
COSMOS	Collating and curating toxicological data for chronic endpoints for cosmetic ingredients	www.cosmostox.eu/
DEMETRA	Data sets for ecotoxicity of pesticides	http://www.demetra-tox.net/index.php
eTOX	Developing a drug safety database from pharmaceutical industry legacy toxicology reports and public toxicology data	www.etoxproject.eu/

Table 5.1 (*Continued*)

Data Resource	Information	Reference
InChemicoTox	DEFRA LINK Alternatives to Animal Testing for Chemical Risk Assessment Project providing curated datasets for *in chemico* assays, skin sensitisation and environmental endpoints	www.inchemicotox.org
OECD (Q)SAR Toolbox	The Toolbox contains a wide range of data on human health and environmental endpoints	http://www.qsartoolbox.org
OSIRIS	Collated data on mutagenicity, carcinogenicity, skin sensitisation, repeat-dose toxicity, bioconcentration factor and aquatic toxicity	www.osiris.ufz.de
ToxCast program	Phase I provided high-throughput screening (HTS) data for over 300 well-characterised chemicals in over 200 assays. Phase II is significantly extending the data generated (approximately 2000 diverse chemicals to be tested using HTS)	http://www.epa.gov/ncct/toxcast/
Tox21 program	Currently screening over 10,000 chemicals at the US National Institute of Health Chemical Genomics Center; Screening approximately 50 chemicals per year using the ToxCast HTS assays	http://epa.gov/ncct/Tox21/
On-line databases		
ChemSpider	Information on over 28 million chemicals (currently) and expanding (includes physico-chemical information, structural information, toxicity in a range of species via different routes of administration)	http://www.chemspider.com/
ChemBL	Database of bioactive drug-like small molecules, containing 2-D structures, calculated properties and abstracted bioactivity data (*e.g.* binding constants, pharmacology and ADMET data)	https://www.ebi.ac.uk/chembl/
ChemIDPlus Advanced	Information on approximately 390,000 chemicals (includes physico-chemical information, structural information, toxicity *etc.*)	http://chem.sis.nlm.nih.gov/chemidplus/
Drugbank	Database for bioinformatics and cheminformatics resource providing detailed drug (*i.e.* chemical, pharmacological and pharmaceutical) data with comprehensive drug target (*i.e.* sequence, structure, and pathway) information; currently contains 6711 drug entries	http://www.drugbank.ca/

Table 5.1 (Continued)

Data Resource	Information	Reference
DSSTox	The Distributed Structure-Searchable Toxicity (DSSTox) Database Network from the US EPA's National Centre for Computational Toxicology provides a forum for publishing downloadable, structure-searchable, chemical structure files associated with chemical inventories or toxicity data sets of environmental relevance	http://www.epa.gov/NCCT/dsstox/
ToxRefDB	Provides results of thousands of *in vivo* toxicity tests on hundreds of chemicals; linked to ToxCast program	
Global (meta) portals		
US EPA ACToR	A data management system that provides links to over 200 databases providing toxicological information	http://actor.epa.gov/actor/faces/ACToRHome.jsp
OECD eChemPortal	Information on properties of chemicals and links to a range of databases providing physico-chemical property, fate and toxicity data	webnet3.oecd.org/echemportal/
ToxNet	Links to many databases providing toxicity information	toxnet.nlm.nih.gov/
Commercial databases		
Leadscope	Data on carcinogenicity, genetic toxicity, chronic and sub-chronic and acute toxicity, reproductive and developmental toxicity	www.leadscope.com
VITIC Nexus	Protocol information and results for carcinogenicity, mutagenicity, genetic toxicity, HERG, hepatotoxicity, skin sensitisation and skin irritation	www.lhasalimited.org
TerraTox	Acute toxicity to environmental species and mammals	www.terrabase-inc.com/
MDL toxicity Database (includes Registry of Toxic Effects of Chemical Substances (RTECS)	Acute toxicity, mutagenicity, skin/eye irritation, tumorigenticity and carcinogenicity, reproductive effects, multiple dose effects data	www.symyx.com
WOMBAT/WOMBAT-PK	Small molecule chemogenomics database; ligand–protein database and clinical pharmacokinetics database	http://www.sunsetmolecular.com/

5.5 Data Quality Assessment

Read-across predictions may rely on data from a very small category or number of compounds to fill a given data gap, for this reason it is imperative that the data used are of high quality. There have been several publications concerning assessment of data quality and fitness-for-purpose.[12–15] This section will summarise some of the key issues identified in this area and tools that are available for assessing data quality.

5.5.1 Accurate Identification and Representation of Chemical Structure

The first issue in collecting any data is to ensure that they are unambiguously associated with the correct chemical — this is also vital for regulatory consideration of predictions (see Section 7.5). This leads to the question of how a given chemical may be identified. There are a range of potential identifiers and ways in which a chemical structure can be represented; these include:

Pictorial representations: there are many ways to represent compounds pictorially in 2 or 3 dimensions; whilst these are often the easiest for humans to interpret they are less useful for storage and input into database search engines.

Nomenclature: common/trivial names, systematic International Union of Pure and Applied Chemistry (IUPAC) names; inventory or database names.

Unique identifiers: in-house identity codes from corporate studies; inventory or database numbers (*e.g.* ChemSpider ID number; European INventory of Existing Commercial chemical substances (EINECS) number); Chemical Abstracts Service (CAS) registry number.

Line notation methods: Simplified Molecular Input Line Entry System (SMILES) strings (including canonicalised SMILES); IUPAC International Chemical Identifier (InChI) code; empirical formulae).

Coding constitutions: these incorporate atom types, coordinates, bond and connectivity information (*e.g.* MDL molfiles) and may have the capacity to include more information (*e.g.* structure data (SD) files) as well as the 3-D conformation of a structure. These can be further hashed into strings of characters (keys) useful for storage and database searching.

Each of the methods of representing a chemical has both advantages and disadvantages as indicated in Table 5.2.

Common problems in the correct identification of chemicals include the use of salt forms in place of parent compounds, isomers/tautomers not being resolved, multiple names and multiple CAS registry numbers existing for an individual compound *etc.* Most of the identifiers listed in Table 5.2 have been available for decades, however, the IUPAC International Chemical Identifier (InChI) was developed more recently as part of an IUPAC project from 2000–2004; development is currently supported via the InChI Trust. InChI was designed to provide an open and standardised format that could be used

Table 5.2 Advantages and disadvantages of different methods for representing chemicals.

Type of representation	Method	Advantages	Disadvantages
Pictorial representations	Many methods available (2D and 3D)	Clearly identifiable structural features; different representations possible for tautomers and isomers	Not acceptable input format for software packages
Nomenclature	Common/trivial names	Generally recognisable structure can be inferred from the name; internationally accepted terminology	Potentially many names for a single compound. Names may be lengthy and complex; may require chemical expertise to interpret
	IUPAC names		
	Inventory/database name (e.g. International Nomenclature for Cosmetic Ingredients (INCI) name)	Applicable within associated inventory/database	Not suitable input for software; requires association with specific inventory/database
Unique identifiers	In-house identity codes	Useful for maintaining confidentiality where access to the data is restricted	Only accessible to a limited number of users; transcription errors possible
	Inventory/database number (e.g. ChemSpider ID number; EINECS number)	Unique identifier	Not suitable input for software; require association with specific inventory/database
	Chemical Abstracts Service (CAS) registry number	Internationally recognised system of identification; a registry number associated with the chemical abstracts database	A single compound may have several CAS registry numbers associated with it (including deleted/referred/alternate/replacing registry numbers); CAS numbers are proprietary and need to be assigned by the American Chemical Society
Line Notations	Chemical formulae	Internationally accepted; interpretable	One formula can represent many compounds; isomers and tautomers are not identified

Table 5.2 (*Continued*)

Type of representation	Method	Advantages	Disadvantages
	SMILES strings	Accepted input format for many chemistry software programs; automated routines for converting structures to SMILES and *vice versa*; interpretable (with experience).	Many SMILES formats can be written for the same compound. Does not provide a unique identifier. Different software interprets SMILES differently.
	Canonical SMILES	Accepted input format for many chemistry software programs; interpretable (with experience).	Canonical SMILES from software packages using different canonical numbering of atoms will not necessarily be comparable.
	InChi Codes	Provides a unique string identifier; interpretable (with experience).	Currently not accepted by many software packages; few automated conversion routines available; non-intuitive in that they require software to convert to a structure that may be interpreted by a human.
Constitution coding	SDfile	Universally accepted; MDL format is used as a standard method; can store compound properties; 3-D structure recorded *e.g.* conformation properties.	Highly format dependent; not useful for database searching.
	Chemistry Mark-up Language (CML)	Generic representation using XML; consistent and software independent; can store annotations and compound properties.	Verbose; not useful for database searching.

Table 5.2 (Continued)

Type of representation	Method	Advantages	Disadvantages
	InChiKeys (Hashed InChi)	Provides a unique identifier in a standard, fixed character length format; useful for storage and database searching (capable of identifying duplicates); short enough to enable internet searches.	Currently not accepted by many software packages; few automated conversion routines available; not readily interpretable by humans.

without restriction, a formalised version of IUPAC names that could be interpreted (with experience) by humans and provide a format useful for searching chemical databases and the internet. The chemical is described in layers of information relating to atoms, bonds, connectivity, tautomeric forms, isotopes, stereochemistry and charge (as appropriate to individual chemicals).

InChIKeys are a condensed version of InChI, comprising 27 characters. They are not interpretable by humans but were designed to make it easier to search for information on the internet. InChI are truly unique, with the condensed InChIKey there is a theoretical, but statistically unlikely, possibility of duplicate codes.

InChiKeys comprise the following format: 14 characters that are a hash of the connectivity information of the InChI; hyphen; 9 characters that are a hash of the remaining layers of the InChI; single character to identify the version of InChI; hyphen; checksum character. An InChI Resolver is required to convert back from the InChIKey to the original InChI.

Table 5.3 gives some of the identifiers possible for paracetamol, showing which are interpretable (by humans) and which require conversion to a more understandable format.

To ensure accuracy of chemical identity, how many identifiers are necessary? Ideally a data set should provide a chemical name, CAS registry number and structure, however few data sets provide all of this information. Commonly a name and a SMILES string or a name and a structure are given. If there is any

Table 5.3 Representations for paracetamol.

Pictorial representation	
Common/trivial names	Paracetamol; acetaminophen; N-(4-hydroxyphenyl) acetanilide; N-(4-hydroxyphenyl) ethanamide; 4-acetamidophenol
IUPAC name	N-(4-Hydroxyphenyl) acetamide
ChemSpider ID	1906
Drugbank ID	DB00316
ChemBL ID	CHEMBL112
EINECS number	203-157-5
CAS registry number	103-90-2
Deleted CAS registry numbers	719293-04-6; 8055-08-1
Chemical Formula	$C_8H_9NO_2$
SMILES string (from ChemIDplus Advanced)	c1(ccc(cc1)O)NC(=O)C
Canonical SMILES (from ChemBL)	CC(=O)Nc1ccc(O)cc1
InChI	InChI=1S/C8H9NO2/c1-6(10)9-7-2-4-8(11)5-3-7/h2-5,11H,1H3,(H,9,10)
InChIKey	RZVAJINKPMORJF-UHFFFAOYSA-N

lack of consistency *e.g.* the structure does not match the given name, then the data cannot be used (unless the error has a readily identifiable cause) as there is no confidence as to which is the correct structure. In some cases further investigation *e.g.* the reference to original data sources to identify transcription errors *etc.* may resolve the problem but correct identification of the chemical associated with a given toxicity is of paramount importance. Certain databases, such as ChemBL (https://www.ebi.ac.uk/chembl/) use data standardisation protocols to ensure accurate representation of chemical structures. Once the correct identity of the chemical has been ascertained the quality of the data that are associated with it can be assessed.

5.5.2 Quality Assessment of Computationally-Derived Chemical Descriptors

In predicting toxicity, useful data can be derived from computational investigation into the properties of the chemicals in question, including the use of quantum chemistry and molecular orbital methods. Physico-chemical descriptors, such as the logarithm of the octanol:water partition coefficient, solubility, volatility *etc.* can be predicted using a range of software.[16] Other useful parameters relate to the steric or electronic properties of the compound such as the electronegativity index, ω, which indicates reactivity of a chemical. When generating and recording data for computationally derived descriptors data quality issues relate to accurate recording of procedures, such as would allow another person to repeat the work. The identity of the chemical must be assured (refer to Section 5.5.1 above), the structure must be entered into the program accurately, *e.g.* misplacement of a substituent on a ring would result in incorrect physico-chemical and structural descriptors being generated. The version number of the program used, default settings, alterations or constraints applied when using the program must be recorded. With the computational chemistry software currently available it is possible to generate thousands of descriptors for a given chemical. A robust system must be used for storing and checking the data. All data and relevant metadata need to be stored in a reliable, portable, system that can be integrated with other software packages as required.

5.5.3 Quality Assessment of Experimentally Derived Data

The term "data quality" can be considered an umbrella term which encompasses many factors. Quality, in terms of correctness of chemical identity and accuracy or reproducibility of computational descriptors of the structure, has been discussed in Sections 5.5.1 and 5.5.2 above. When assessing experimental data many more aspects that contribute to overall data quality need to be considered. This has given rise to various definitions and schemes used to describe data quality. Klimisch *et al.* devised a systematic approach to evaluate toxicological data quality, which has become one of the most widely accepted methods of quality assessment, (discussed further in Section 5.5.4).[17] The authors proposed three aspects contributing to overall data quality; reliability, relevance and adequacy. These and other terms commonly used to describe data quality are discussed below:

Reliability was defined by Klimisch *et al.* as "evaluating the *inherent quality* of a test report or publication relating to preferably standardised methodology and the way the experimental procedure and results are described to give evidence of the clarity and plausibility of the findings."[17]

For a true assessment of data quality, ideally the experimental protocol should be scrutinised to ensure the procedures were sufficiently robust to provide reliable results. Consider the testing of a chemical in an *in chemico* or an *in vitro* assay, factors that may impact on the reliability of the results include:

- experience of the personnel/laboratory performing the test;
- adherence to guidelines (*i.e.* are there standardised protocols for the assay and is any deviation from these clearly justified?; is Good laboratory Practice (GLP) observed?);
- compound identity (*i.e.* parent or salt form used);
- compound purity;
- vehicle or solubilising agents used;
- susceptibility to oxidation, hydrolysis or other abiotic transformation;
- solubility in test media;
- volatility;
- statistical method used to analyse the results and the selection of positive and negative controls (error ranges due to technical and/or organism variability should be recorded, raw data made available where possible).

For *in vivo* assays many of the above are also relevant but there are additional considerations, such as:

- confirmation of the dose administered and the route of administration;
- sample size (use of pooled data or individual data);
- environmental considerations (housing, feeding, water availability, *etc.*);
- potential biological transformation (metabolism) of compound, particularly where the metabolite(s) may contribute to the toxic effect.

The above lists some sources of variability that may affect the outcome of an *in vivo* assay, however there are many other influential factors discussed in more detail by Nendza *et al.* and Madden *et al.*[12,18] When collating toxicity data it is frequently the case that data are collected from a wide range of sources. This can lead to inter-laboratory variation in procedures leading to variability in results. Detailed reporting of protocols can help to identify where variability may have influenced the toxicity values reported. Data that are evaluated as being less reliable can still be useful. For example, where several "low quality" studies have all shown the same result; the result from one of these would not be reliable on its own but combined with the evidence from other studies then weight-of-evidence would support the result.

Data reliability is inherent in the procedure *i.e.* it is independent of purpose of the data, relevance and adequacy are however dependent on the intended use of the data.

Relevance was defined by Klimisch *et al.* as "covering the extent to which data and/or tests are appropriate for particular hazard identification or risk characterisation."[17] Another commonly used term is *fitness-for-purpose i.e.* are the data suitable for the intended purpose? A great deal of toxicological data have been generated over decades. Although not specifically generated for the purpose of category formation or the building of *in silico* models, these data can be assessed to determine their suitability for these purposes. In terms of risk assessment, the EChA guidance (Chapter R.4) on evaluating relevance of data for REACH, suggests consideration of the following:[14]

- Was the substance tested representative for the substance as being registered?
- Has the appropriate species been chosen?
- Is the route of exposure relevant for the population?
- Were appropriate doses/concentrations tested?
- Were the critical parameters influencing the endpoint considered adequately?

Human data are the most relevant, but data from other species can be useful if its relevance to human toxicity can be established.

Adequacy: was defined by Klimisch *et al.* as "defining the usefulness of data for hazard/risk assessment purposes.[17] When there is more than one set of data for each effect, the greatest weight is attached to the most reliable and relevant." Many individual measures of toxicity can be associated with a particular toxic event within an organism. For example, measurements of specific liver enzyme levels may be useful indicators for hepatotoxicity. Zweers and Vermeire provide a list of factors, similar to those considered above, for determining reliability and relevance.[19] The authors conclude that expert judgement is the most important factor in determining relevance and adequacy. A clear rationale must be given when justifying the adequacy of data for a given purpose.

Accuracy is a key term applicable to every step of data generation from accuracy of experimental measurement to accurate and complete recording of data. OECD GD 34 defines accuracy as "the closeness of agreement between test method results and accepted reference values.[15] It is a measure of test method performance and one aspect of relevance. The term is often used interchangeably with "concordance" to mean the proportion of correct outcomes of a test method." The accuracy of a measurement is inversely linked to uncertainty. *Uncertainty* was defined by Nendza *et al.* as "the estimated amount or percentage (margin of error) by which an observed or calculated value obtained by a method may differ from the true value.[12] Uncertainty depends on both accuracy and precision and is caused by either variability or lack of knowledge."

Completeness is another aspect of quality; how many entries must be recorded for a single compound (and its associated data) for the record to be considered "complete"? The answer to this depends entirely on the nature of the investigation and there are vast possibilities for items that may be considered important depending on the nature of the study (examples include structure, name, CAS, InChI, SMILES, program version and settings, dose, dose–response curve, endpoint toxicity value, timing of measurements, route of administration, species, strain, sex, age, housing, bedding, cell-type, presence of metabolising enzymes in *in vitro* systems, vehicle, solubilising agents *etc.*). Each of these may induce variability in experimental results but in collecting data a pragmatic approach should be adopted in terms of what is a complete record and what is an acceptable level of information.[18] Checklist approaches have been proposed previously and initiatives such as the ARRIVE guidelines (Animals in Research: Reporting *In Vivo* experiments) have been published recently with the aim of standardising reporting of experimental work in future publications.[12,20] Completeness of data may have more formal connotations for example what is considered "complete" information in terms of a REACH submission for regulatory purposes. In these cases the data requirements must be clearly stipulated. Fu *et al.* define a complete data source as one which "covers adequate data in both depth and breadth to meet the defined business information demand."[21]

Reproducibility can be assessed between laboratories, or within the same laboratory. Inter-laboratory reproducibility is defined in OECD GD 34 as "a measure of the extent to which different qualified laboratories, using the same protocol and testing the same substances, can produce qualitatively and quantitatively similar results.[15] It indicates the extent to which a test can be successfully transferred between laboratories. Intra-laboratory *repeatability* is the closeness of agreement between test results obtained within a single laboratory when the procedure is performed on the same substance under identical conditions within a given time period. Well-designed experiments with a clear and detailed protocol should be able to generate highly reproducible and repeatable results.

Validation is defined in OECD GD 34 as the process by which the reliability and relevance of a particular approach, method, process or assessment is established for a defined purpose. A *valid* test method is one with sufficient relevance and reliability for a specific purpose and which is based on scientifically sound principles. A test method is never valid in an absolute sense, but only in relation to a defined purpose. The "Solna Principles", developed at an OECD Workshop in 1996, agreed the following principles as applied to the validation of new or updated test methods for hazard assessment:[15]

(1) A rationale for the test method should be available. This should include a clear statement of scientific need and regulatory purpose.

(2) The relationship of the endpoint(s) determined by the test method to the *in vivo* biological effect and to the toxicity of interest should be addressed. The limitations of a method should be described, *e.g.*, metabolic capability.
(3) A formal detailed protocol must be provided and should be readily available in the public domain. It should be sufficiently detailed to enable the user to adhere to it, and it should include data analysis and decision criteria. Test methods and results should be available preferably in an independent peer reviewed publication. In addition, the result of the test should have been subjected to independent scientific review.
(4) Intra-test variability, repeatability and reproducibility of the test method within and amongst laboratories should have been demonstrated. Data should be provided describing the level of inter- and intra-laboratory variability and how these vary with time.
(5) The test method's performance must have been demonstrated using a series of reference chemicals preferably coded to exclude bias.
(6) The performance of test methods should have been evaluated in relation to existing relevant toxicity data as well as information from the relevant target species.
(7) All data supporting the assessment of the validity of the test methods including the full data set collected in the validation study must be available for review.
(8) Normally, these data should have been obtained in accordance with the OECD Principles of Good Laboratory Practice (GLP).

5.5.4 Guidance and Tools for Data Quality Assessment

Assessing data quality is now recognised as a major factor in developing high quality, predictions with greater acceptability. In this section quality assessment schemes and available tools will be discussed.

The Klimisch scheme has become one of the most widely accepted schemes for assessing data quality.[17] The Klimisch scheme was devised for evaluating quality of toxicological and ecotoxicological data but can also be applied to other areas. Application of the scheme enables data to be placed in one of four categories for reliability; these are (1) reliable without restriction (2) reliable with restrictions (3) not reliable and (4) not assignable. These categories were defined as shown in Table 5.4

The definitions for the categories given in Table 5.4 provide general guidance on classifying data, however, more specific guidance is needed on how to evaluate the available information in order to place data into one of these four categories. Klimisch *et al.* list the factors that should be recorded and evaluated for tests not carried out in accordance with international guidelines. Consideration of these factors should aid the decision as to how to categorise the data. In an attempt to formalise the consideration of these

Table 5.4 Definitions of the categories devised by Klimisch et al.[17]

Code	Category
1	**Reliable without restriction** "Studies or data from the literature or reports which were carried out or generated according to generally valid and/or internationally accepted testing guidelines (preferably performed according to GLP) or in which the test parameters documented are based on a specific (national) testing guideline (preferably performed according to GLP) or in which all parameters described are closely related/comparable to a method."
2	**Reliable with restrictions** "Studies or data from the literature, reports (mostly not performed according to GLP), in which the test parameters documented do not totally comply with the specific testing guideline, but are sufficient to accept the data or in which investigations are described which cannot be subsumed under a testing guideline, but which are nevertheless well documented and scientifically acceptable."
3	**Not reliable** "Studies or data from the literature/reports in which there were interferences between the measuring system and the test substance or in which organisms/test systems were used which are not relevant in relation to the exposure (*e.g.*, unphysiologic pathways of application) or which were carried out or generated according to a method which is not acceptable, the documentation of which is not sufficient for assessment and which is not convincing for an expert judgment."
4	**Not assignable** "Studies or data from the literature, which do not give sufficient experimental details and which are only listed in short abstracts or secondary literature (books, reviews, *etc.*)."

factors, and so make data evaluation more consistent between researchers, the Toxicological data Reliability Assessment Tool (ToxRTool) was developed as part of a project funded by the European Centre for the Validation of Alternative Methods (ECVAM). The aim of the project was to provide a transparent tool, for scientists routinely involved in data reliability assessment, which would result in greater consistency in reliability scores for data. Details of ToxRTool and its evaluation are given in Schneider *et al.*[22] The tool comprises two key spreadsheets (one for *in vitro* data and one for *in vivo* data). The questions within the spreadsheets are based on issues that Klimisch *et al.* recommended for consideration and are divided into five criteria groups. Tables 5.5 and 5.6 show the questions posed for the assessment of *in vivo* data and *in vitro* data respectively.

Answering yes to any question in the spreadsheet scores 1 point. Hence the scores for the assessment of *in vivo* data range from 0 to 21 and for *in vitro* data range from 0 to 18. The overall scores are converted into a Klimisch category as shown in Table 5.7. Note that certain questions in Tables 5.5 and 5.6 are annotated with a "*"; these are referred to as "red" criteria questions. For strict assessment of quality, failure to answer yes to any of these "red" (or essential) criteria will automatically result in a Klimisch code of 3 (not reliable) being assigned to the data. Within ToxRTool two outputs are possible: an original score which discounts red criteria (*i.e.* answering no to one of the

Table 5.5 ToxRTool questions posed for the assessment of *in vivo* data.[22]

No.	Criteria Group I: Test substance identification	Score
*1	Was the test substance identified?	
2	Is the purity of the substance given?	
3	Is information on the source/origin of the substance given?	
4	Is all information on the nature and/or physico-chemical properties of the test item given, which you deem indispensable for judging the data?[a]	
	Criteria Group II: Test organism characterisation	
*5	Is the species given?	
6	Is the sex of the test organism given?	
7	Is information given on the strain of test animals plus, if considered necessary to judge the study, other specifications?[a]	
8	Is age or body weight of the test organisms at the start of the study given?	
9	For repeated dose toxicity studies only (give point for other study types): Is information given on the housing or feeding conditions?	
	Criteria Group III: Study design description	
*10	Is the administration route given?	
*11	Are doses administered or concentrations in application media given?	
*12	Are frequency and duration of exposure as well as time-points of observations explained?	
*13	Were negative (where required) and positive controls (where required) included? (give point also, when absent but not required)[a]	
*14	Is the number of animals (in case of experimental human studies: number of test persons) per group given?	
15	Are sufficient details of the administration scheme given to judge the study?[a]	
16	For inhalation studies and repeated dose toxicity studies only (give point for other study types): Were achieved concentrations analytically verified or was stability of the test substance otherwise ensured or made plausible?	
	Criteria Group IV: Study results documentation	
17	Are the study endpoint(s) and their method(s) of determination clearly described?	
18	Is the description of the study results for all endpoints investigated transparent and complete?	
19	Are the statistical methods applied for data analysis given and applied in a transparent manner? (give also point, if not necessary/applicable)[a]	
	Criteria Group V: Plausibility of study design and results	
*20	Is the study design chosen appropriate for obtaining the substance-specific data aimed at?[a]	
21	Are the quantitative study results reliable?[a]	

[a]Refers to points in the spreadsheet where additional information is provided as supporting notes to aid the user in making a decision.
*These questions are referred to as "red" or essential criteria (refer to text).

Table 5.6 ToxRTool questions posed for the assessment of *in vitro* data.[22]

No.	Criteria Group I: Test substance identification	Score
*1	Was the test substance identified?	
2	Is the purity of the substance given?	
3	Is information on the source/origin of the substance given?	
4	Is all information on the nature and/or physico-chemical properties of the test item given, which you deem <u>indispensable</u> for judging the data?[a]	
	Criteria Group II: Test system characterisation	
5	Is the test system described?	
6	Is information given on the source/origin of the test system?	
7	Are necessary information on test system properties, and on conditions of cultivation and maintenance given?	
	Criteria Group III: Study design description	
8	Is the method of administration given (see explanations for details)?	
*9	Are doses administered or concentrations in application media given?	
*10	Are frequency and duration of exposure as well as time-points of observations explained?	
*11	Were negative controls included? (give also point, if not necessary)[a]	
*12	Were positive controls included? (give also point, if not necessary)[a]	
13	Is the number of replicates (or complete repetitions of experiment) given?	
	Criteria Group IV: Study results documentation	
14	Are the study endpoint(s) and their method(s) of determination clearly described?	
15	Is the description of the study results for all endpoints investigated transparent and complete?	
16	Are the statistical methods for data analysis given and applied in a transparent manner? (give also point, if not necessary/applicable)[a]	
	Criteria Group V: Plausibility of study design and data	
*17	Is the study design chosen appropriate for obtaining the substance-specific data aimed at?[a]	
18	Are the <u>quantitative</u> study results reliable?[a]	

[a]Refers to points in the spreadsheet where additional information is provided as supporting notes to aid the user in making a decision.
*These questions are referred to as "red" or essential criteria (refer to text).

highlighted question will score 0 but will not automatically place the data into category 3; and a revised score (in which red criteria are considered).

Using the ToxRTool automatically results in assignation to categories 1, 2 or 3. It is not possible for data to be classified into category 4 (not assignable) using ToxRTool. The tool is freely downloadable from the JRC website (http://ecvam.jrc.it; section 'Publications').

Table 5.7 Output of the ToxRTool in terms of Klimisch codes.

Score for in vivo data	Score for in vitro data	Associated Klimisch Code	Inference
18–21	15–18	1 - reliable without restriction	Useful, check relevance for intended purpose
13–17	11–14	2 - reliable with restrictions	Potentially useful, check relevance for intended purpose
<13 or not all *red* criteria met	<11 or not all *red* criteria met	3 - not reliable	Generally not to be used as a key study, but depending on the shortcomings of the study it may still be useful in weight-of-evidence approaches or as supportive information
		4 - not assignable (note use of ToxRTool cannot give a classification of Klimisch Code 4)	Generally not to be used as key study, but depending on the shortcomings of the study it may still be useful in weight-of-evidence approaches or as supportive information.

5.5.5 Alternative Assessment Schemes

Whilst the above provides a definitive, quantitative approach to data quality assessment, more qualitative approaches may also be useful. Przybylak *et al.* reviewed chemical and biological factors relevant to quality assessment and proposed four criteria for consideration:

(1) availability of sufficient supporting information; data published in study report or peer review paper; reasonably easy access to original reference.
(2) consistency of study design; the compliance with internationally accepted guidelines (*e.g.* OECD, US EPA).
(3) GLP; compliance with principles of GLP.
(4) identification of test material; correct chemical name, CAS number and 2D/3D structure, information about purity and source, physico-chemical characterisation, when required.

Each of these four criteria can give rise to additional queries. For example: (1) The supporting information must not only be available, it also needs to be correct and sufficiently detailed. Peer review papers may be available but journals differ in their information requirements for published studies. This is issue was addressed by Kilkenny *et al.* in proposing the ARRIVE guidelines for the publication of experimental studies.[20] The guidelines list 20 essential pieces of information that should be reported for all studies). (2) Consistent study design allows for easier comparison of results between laboratories, reducing variability in results. Studies performed prior to guidelines being published can also be useful. (3) Whilst modern studies are generally GLP compliant, GLP refers to adherence to protocols and reporting, issues of lack of sensitivity/relevance of the test are not identified by adherence to GLP.

Other data quality assessment schemes have been published for more specific purposes, but are not discussed here. The reader is referred to the given references for more information: (Tielemans *et al.* for the evaluation of exposure data; Hobbs *et al.* for the evaluation of aquatic toxicity data; US EPA for assessing adequacy of data for the High Production Volume Challenge program.).[23-25]

5.5.6 Problems with Assessment

One key item in data quality assessment is whether assessment is carried out for the data set as a whole or whether assessment is made for each individual chemical. This issue of global versus chemical-specific assessment is discussed in more detail by Przybylak *et al.*[13] Within a data set generally considered to be of high quality, there may be certain data points of low quality. This can arise for example if the compound has low solubility in the vehicle or has a high volatility and is used in an open test system. In such cases the true concentration will be less than the nominal concentration giving a spurious result for potency. In many modelling exercises it is not reasonably practicable to assess each individual datum point for quality. However, for read-across, where very few compounds may be present in the final category then this may become possible. This gives higher confidence to the data in the category, hence higher confidence in the prediction.

Schneider *et al.*, in evaluating the ToxRTool, identified several problems in trying to devise a unified approach to data quality assessment.[22] As part of the evaluation a panel of assessors were asked to quality assess data from the same publications using the ToxRTool (and prototype versions). Problems arose due to assessors interpreting information given in the publications differently, failing to note important information that was provided in the publication and interpreting the quality criteria differently. Some assessors adhered more strictly to the guidelines and others had a more flexible approach. Such differences of opinion are difficult to resolve given the different background, experience and expectations of the assessors. In an attempt to increase flexibility and transparency in assigning quality criteria scores to data, Yang *et al.* developed an experimental fuzzy expert system, based on ToxRTool, for assessing the quality of toxicological data.[26]

Agerstrand *et al.* compared four methods for evaluating reliability of nine non-standard ecotoxicity data sets.[27] Their findings showed that the different methods gave different outcomes for the overall evaluation of quality for seven of the nine datasets investigated. They concluded that methods to reduce "vagueness and elements of case-by-case interpretations" should be sought but that expert judgement should remain a part of the evaluation process.

A problem of trying to implement standardised criteria for reporting data at the present time, to make quality assessment of this data more consistent in future, is difficult. This is because it would require a pre-judgement of what quality criteria would be dependent upon in the future.[28] What determines "quality" in terms of fitness-for-purpose is dependent upon what the purpose is. For example, data may be high quality for one purpose (sufficiently detailed for ranking) but low quality for another purpose (not sufficiently accurate for read-across). Hence it is difficult to give a particular quality rating to a data set. One useful consideration is "corroborating evidence" *i.e.* have other researchers assessed the quality of a data set for a given purpose. Fu *et al.* discussed the potential for a repository of information concerning available data sets.[21] This could include what other researchers have used the data for, how they assessed its quality and if there were any issues identified with the data set as a whole or individual datum points within the data set. This could provide useful information for subsequent modellers, preventing duplication of effort in terms of performing a quality assessment and would provide a source of information where problems have been identified.

Conflicting data present a real problem in any assessment of quality. Studies that have been performed to a high standard can produce conflicting results, such is the nature of biological systems in particular. In cases where conflicting results have been identified then resolution is required before the data can be used. This may require further, more detailed investigation into the studies to identify potential sources of variation, for example in *in vitro* systems, conflicting results may arise depending on whether or not metabolising enzymes were incorporated in the systems. Where an obvious reason for discrepancy cannot be found then weight-of-evidence may be used *i.e.* confirmation is sought from other studies for the same compound. If the conflict cannot be resolved then the data cannot be used. It is important to note that a lack of supporting information, or lack of information leading to a classification of not assignable in the Klimisch scheme, does not mean data are low quality, it merely means information is not available on which to make a judgement. If more information can be obtained on a study of interest then it may become possible to assign the data to a category and so have confidence in its use.

5.6 Conclusions

The importance of data quality assessment is to ascertain the suitability of the data for a given purpose. In terms of read-across a range of data may be useful and these need to be of high quality. Data relating to a range of characteristics of chemicals may be relevant; these include physico-chemical, structural or reactivity properties in addition to biological activity, or toxicity data. This enables reliable predictions for compounds of unknown activity to be inferred from other category members of known activity.

No data are ideal; there will always be a level of uncertainty associated with them. What is important is that the sources of uncertainty can be identified and communicated, such that the use of particular data is justifiable. Whilst high quality data are preferable, lower quality data may also be useful if combined with other data in a weight-of-evidence approach. Best use must be made of available data, particularly where data for a given endpoint are scarce. Should new data become available it should be used for confirmatory purposes to increase confidence in predictions.

Several problems in data quality assessment have been identified in this chapter showing that none of the current assessment schemes are ideal. Irrespective of which scheme is used to assess quality, what is important is that some form of quality assessment is performed that justifies use of the data and so gives greater confidence to read-across predictions.

Acknowledgements

The funding from the European Community's Seventh Framework Program (FP7/2007-2013) COSMOS Project under grant agreement no. 266835 and from Cosmetics Europe and the funding of the IMI-JU eTOX prject (grant agreement no. 115002) are gratefully acknowledged.

References

1. V. T. Politano and A. M. Api, The Research Institute for Fragrance Materials human repeated insult patch test protocol, *Regul. Toxicol. Pharmacol.*, 2008, **52**, 35.
2. Organisation for Economic Co-operation and Development (OECD), 1992, Guidelines for Testing of Chemicals, *No. 406. Skin Sensitisation*.
3. Organisation for Economic Co-operation and Development (OECD), 2004 Guidelines for Testing of Chemicals, *No. 429. Skin Sensitisation: Local Lymph Node Assay*.
4. G. Maxwell, P. Aeby, T. Ashikaga, S. Bessou-Touya, W. Diembeck, F. Gerberick, P. Kern, M. Marrec-Fairley, J. M. Ovigne, H. Sakaguchi, K. Schroeder, M. Tailhardat, S. Teissier, and P. Winkler, Skin Sensitisation: The Colipa strategy for developing and evaluating non-animal test methods for risk assessment, *Altex*, 2011, **28**, 50.
5. R. J. Kavlock, D. J. Dix, K. A. Houck, R. S. Judson, M. T. Martin and A. Richard, ToxCast: Developing predictive signatures for chemical toxicity *AATEX (Special Issue Proceedings of the 6th World Congress on Alternatives and Animal Use in the Life Sciences)*, 2007, **14**, 623.
6. M. T. D. Cronin, Finding the Data to Develop and Evaluate (Q)SARs and Populate Categories for Toxicity Prediction in *In Silico Toxicology: Principles and Applications*, ed. M. T. D. Cronin and J. C. Madden, The Royal Society of Chemistry, Cambridge, 2010, p. 31.
7. J. A. H. Schwöbel, Y. K. Koleva, S. J. Enoch, F. Bajot, M. Hewitt, J. C. Madden, D. W. Roberts, T. W. Schultz, and M. T. D. Cronin, Measurement and estimation of electrophilic reactivity for predictive toxicology, *Chem. Rev.*, 2011, **111**, 2562

8. A. S. Kalgutkar, I. Gardner, R. S. Obach, C. L. Shaffer, E. Callegari, K. R. Henne, A. E. Mutlib, D. K. Dalvie, J. S. Lee, Y. Nakai, J. P. O'Donnell, J. Boer and S. P. Harriman, A comprehensive listing of bioactivation pathways of organic functional groups, *Curr. Drug Metab.*, 2005, **6**, 161.
9. D. Fourches, J. C. Barnes, N. C. Day, P. Bradley, J. Z. Reed, and A. Tropsha, Cheminformatics analysis of assertions mined from literature that describe drug-induced liver injury in different species, *Chem. Res. Tox.*, 2010, **23**, 171.
10. S. J. Enoch, C. M. Ellison, T. W. Schultz, and M. T. D. Cronin, A review of the electrophilic reaction chemistry involved in covalent protein binding relevant to toxicity, *Crit. Rev. Toxicol.*, 2011, **41**, 783.
11. S. J. Enoch and M. T. D. Cronin, Development of new structural alerts suitable for chemical category formation for assigning covalent and non-covalent mechanisms relevant to DNA binding, *Mutat. Res.–Genetic Toxicol. Environ. Mutagen*, 2012, **743**, 10.
12. M. Nendza, T. Aldenberg, E. Benfenati, R. Benigni, M. Cronin, S. Escher, A. Fernandez, S. Gabbert, F. Giralt, M. Hewitt, M. Hrovat, S. Jeram, D. Kroese, J. C. Madden, I. Mangelsdorf, R. Rallo, A. Roncaglioni, E. Rorije, H. Segner, B. Simon-Hettich and T. Vermeire, data quality assessment for in silico methods: a survey of approaches and needs in *in silico toxicology: principles and applications*, ed. M. T. D. Cronin and J. C. Madden, The Royal Society of Chemistry, Cambridge, 2010, p. 59.
13. K. R. Przybylak, J. C. Madden, M. T. D. Cronin and M. Hewitt, Assessing toxicological data quality: basic principles existing schemes and current limitations, *SAR QSAR Env. Res.*, 2012, **23**, 435.
14. European Chemicals Agency (ECHA) Guidance on information requirements and chemical safety assessment Chapter R.4: Evaluation of available information. Available at: http://echa.europa.eu/documents/10162/13643/information_requirements_r4_en.pdf (Accessed April 2013).
15. Organisation for Economic Cooperation and Development (OECD) Series on Testing and Assessment, Number 34: *Guidance document on the validation and international acceptance of new or updated test methods for hazard assessment* ENV/JM/MONO, 2005, 14 Available from http://www.oecd.org/officialdocuments/displaydocumentpdf/?cote=env/jm/mono(2005)14&doclanguage=en (Accessed April 2013).
16. T. H. Webb and L. Morlacci, calculation of physico-chemical and environmental fate properties, in *in silico toxicology: principles and applications*, ed. M. T. D. Cronin and J. C. Madden, The Royal Society of Chemistry, Cambridge, 2010, p. 118.
17. H. -J. Klimisch, M. Andreae and U. Tillmann, A systematic approach for evaluating the quality of experimental toxicological and ecotoxicological data, *Regul. Toxicol. Pharmacol.* 1997, **25**, 1.
18. J. C. Madden, M. Hewitt, K. Przybylak, R. J. Vandebriel and A. H. Piersma, Strategies for the optimisation of in vivo experiments in

accordance with the 3Rs philosophy, *Regul. Toxicol. Pharmacol.*, 2012, **63**, 140.
19. P. G. P. C. Zweers and T. G. Vermeire, Data: Needs, Availability, Sources and Evaluation in *Risk Assessment of Chemicals*: *An Introduction* (2nd Ed), ed C. J. van Leeuwen and T. G. Vermeire, Springer, The Netherlands, 2007, p. 357.
20. C. Kilkenny, W. J. Browne, I. C. Cuthill, M. Emerson and D. G. Altman, Improving bioscience research reporting: the ARRIVE guidelines for reporting animal research, *PLoS Biol.* 2010, **8**, e1000412.
21. X. Fu, A. Wojak, D. Neagu, M. Ridley and K. Travis, Data governance in predictive toxicology: A review, *J. Cheminformatics.*, 2011, **3**, 24.
22. K. Schneider, M. Schwarz, I. Burkholder, A. Kopp-Schneider, L. Edler, A. Kinsner-Ovaskainen, T. Hartung and S. Hoffmann S, "ToxRTool", a new tool to assess the reliability of toxicological data, *Toxicol. Lett.*, 2009, **189**, 138.
23. E. Tielemans, H. Marquart, J. De Cock, M. Groenewold and J. Van Hemmen, A proposal for evaluation of exposure data, *Ann. Occup. Hyg*, 2002, **46**, 287.
24. D. A. Hobbs, M. S. Warne and S. J. Markich, Evaluation of criteria used to assess the quality of aquatic toxicity data. *Integr. Environ. Assess. Manage.*, 2005, **1**, 174.
25. US EPA, Determining the adequacy of existing data, guideline for the HPV challenge program, Available at http://www.epa.gov/hpv/pubs/general/datadfin.htm (Accessed April 2013).
26. L. Yang, D. Neagu, M. T. D. Cronin, M. Hewitt, S. J. Enoch, J. C. Madden and K. Przybylak, Towards a fuzzy expert system on toxicological data quality assessment, *Mol. Inf*, 2013, **32**, 65.
27. M. Agerstrand, M. Breitholtz and C. Ruden, Comparison of four different methods for reliability evaluation of ecotoxicity data: a case study of non-standard test data used in environmental risk assessments of pharmaceutical substances, *Environ. Sci. Eur.*, 2011, **23**, 17.
28. P. N. Judson, P. A. Cooke, N. G. Doerrer, N. Greene, R. P. Hanzlik, C. H. Hardy, A. Hartmann, D. Hinchliffe, J. Holder, L. Muller, T. Steger-Hartmann, A. Rothfuss, M. Smith, K. Thomas, J. D. Vessey and E. Zeiger, Towards the creation of an international toxicology information centre, *Toxicology*, 2005, **213**, 117.

CHAPTER 6
Category Formation Case Studies

S. J. ENOCH*, K. R. PRZYBYLAK AND M. T. D. CRONIN

Liverpool John Moores University, School of Pharmacy and Chemistry, Byrom Street, Liverpool, L3 3AF, England
*E-mail: s.j.enoch@ljmu.ac.uk

6.1 Introduction

The aim of this chapter is to outline a number of case studies for category formation using freely available toxicological data. The information presented is not intended as a guidance document or to be an exhaustive review of the published literature. Instead, the chapter aims to outline examples of how knowledge of the molecular initiating event, incorporated into profilers, can be used to develop chemical categories. Thus, the main focus relates to the use of the information contained within the OECD QSAR Toolbox (for more information see Section 4.3). This tool is freely available, contains many toxicological data and a variety of profilers. It is envisaged that information relating to Adverse Outcome Pathways (refer to Chapter 3) will be incorporated into it as the field advances. The chapter concludes with a case study based on the use of structural similarity to form categories. This approach, whilst less common, is still useful, especially for endpoints where little or no mechanistic information exists with which to develop suitable profilers.

6.2 Mechanism-based Case Studies

The following section outlines how information related to the molecular initiating event (MIE) can be used to group chemicals into categories allowing data gaps to be filled using read-across. This mechanistic information has been incorporated into the OECD QSAR Toolbox[1] (version 3.1 at the time of writing, refer to Section 4.3 for more information) and this section will outline the use of profilers within the OECD QSAR Toolbox to develop chemical categories suitable for data gap filling for a number of endpoints. The OECD QSAR Toolbox enables chemical categories to be formed using *in silico* profilers that consist of structural alerts. There are two types of profiler that are used to develop categories; primary and secondary profilers.

The primary profilers are sub-divided into either mechanistic or endpoint specific profilers. Mechanistic primary profilers contain structural alerts that have been developed around the chemistry related to a specific molecular initiating event (for example, covalent bond formation between a chemical and a protein). The structural alerts within this type of profiler are not necessarily supported by toxicological data. In contrast, the endpoint specific profilers contain structural alerts that have been identified from analysis of toxicological data. The mechanistic and endpoint-specific primary profilers are complementary, with the ideal scenario being that a mechanistic profiler identifies a single molecular initiating event that is supported by a structural alert identified by an appropriate endpoint specific profiler. It is important to realise that the mechanistic profilers offer a broad coverage of the chemical space related to a molecular initiating event. In contrast, the endpoint specific profilers are focused on a narrow area of chemical space for which toxicological data exist. This can be considered in terms of a general Adverse Outcome Pathway (AOP) with the mechanistic profilers relating to the molecular initiating event. The endpoint specific profilers then help to focus the category towards the endpoint of interest *i.e.* they help anchor the category with the aid of information derived from toxicological data (see Chapter 3).

The secondary profilers contain structural alerts that enable simple features to be identified in chemicals within a category. The two most useful secondary profilers enable chemical elements and organic functional groups to be identified. This type of profiling enables the structural domain of a category, developed using the primary profilers, to be defined. This is important as defining the structural domain in terms of the target chemical (the chemical around which the category is being developed) helps to ensure confidence in the resulting predictions.

6.2.1 Case Study One: Category Formation for Ames Mutagenicity

This case study outlines the formation of a chemical category suitable for read-across for mutagenicity as measured in the Ames assay. This assay uses a number of genetically engineered strains of *Salmonella* to detect chemicals capable of causing damage to DNA by one of two molecular initiating events; covalent adduct formation or intercalation with DNA.[2] The case study outlines category formation for chemicals capable of forming a covalent bond with DNA using 3-methylaniline as the target chemical.

Step 1: Initial profiling using the primary profilers

Profiling the 3-methylaniline with the two mechanistic profilers related to covalent bond formation with DNA ('DNA binding by OECD' and 'DNA binding by OASIS') showed 3-methylaniline to contain a primary aromatic amine moiety capable of an S_N1 reaction via the formation of a nitrenium ion (Figure 6.1). The mechanistic profiling results were supported by the identification of an aromatic amine moiety by the endpoint specific profiler ('*in vitro* mutagenicity (Ames test) by ISS'). The complementary nature of the results between the mechanistic profilers and endpoint specific profiler was taken as an indication of confidence in the profiling results.

Step 2: Initial category formation and sub-categorisation using the primary profilers

The initial profiling results were then used to retrieve mechanistic analogues from the applicable databases within the OECD QSAR Toolbox. The relevant databases that were searched were:

- Bacterial mutagenicity ISSSTY
- Genotoxicity OASIS

Profiling these databases with the 'DNA binding by OECD' profiler created an initial category of 554 chemicals (including the target chemical). This category required sub-categorisation in order to ensure that it contains only analogues that act via the same mechanism as 3-methylaniline. Thus, the following sub-categorisations were required:

- Sub-categorisation of the initial category of 554 chemicals using the 'DNA binding by OECD' profiler (the profiler that was used to develop the initial category). This sub-categorisation resulted in a category of 317 chemicals.
- Sub-categorisation of the category of 317 chemicals using the 'DNA binding by OASIS' profiler. This sub-categorisation resulted in a category of 293 chemicals.

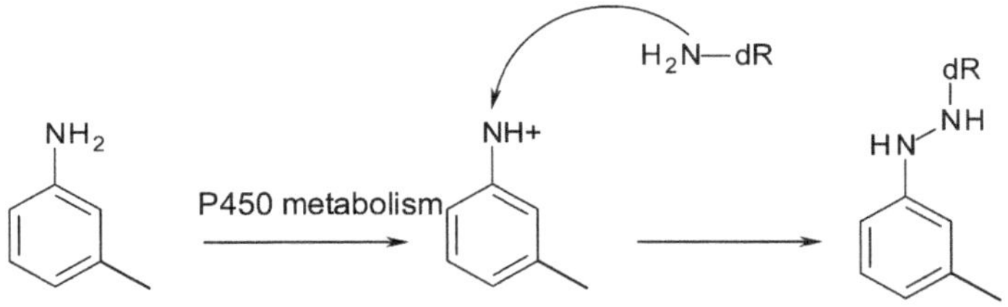

Figure 6.1 Electrophilic mechanism for 3-methylaniline (dR = DNA chain).

- Sub-categorisation of the category of 293 chemicals using the '*in vitro* mutagenicity (Ames test) by ISS' profiler. This sub-categorisation resulted in a category of 234 chemicals.

Step 3: Empiric sub-categorisation using the secondary profilers

The chemical category defined in Step 2 using the primary profilers contained a wide range of chemicals with differing functional groups compared to those of the target chemical. Thus, it was necessary to define the structural domain of the category in terms of the target chemical. This involved a series of sub-categorisations using two of the empiric secondary profilers as follows:

- Sub-categorisation of the category of 234 chemicals using the 'chemical elements' profiler. This sub-categorisation resulted in a category of 85 chemicals.
- Sub-categorisation of the category of 85 chemicals using the 'organic functional group' profiler. This sub-categorisation in a final category of 10 unique chemicals (including the target chemical).

Step 4: Data-gap filling via read-across

The sub-categorisation carried out using the primary and secondary profilers resulted in a category that had a well-defined mechanistic and structural domain. This category was then used to fill the hypothetical data gap for 3-methylaniline in the TA1537 strain of *Salmonella* in the presence of the S9 liver fraction. Inspection of the category showed that only seven chemicals had associated toxicological data (Table 6.1). These data were used to make a read-across prediction that 3-methylaniline would be negative if tested in the TA 1537 strain of *Salmonella*.

The results of the case study showed that the following approach can be considered as a good method for the development of chemical categories for mutagenicity.

- Profile the target chemical to identify a potential MIE using mechanistic profilers related to covalent DNA binding and an endpoint specific profiler related to mutagenicity. Confidence is gained in the profiling results if the mechanistic and endpoint specific profilers are in agreement.
- Identify chemicals capable of eliciting the same MIE as the target chemical, from suitable toxicological databases. This can be achieved using the mechanistic profilers utilised in Step 1. The resulting category is termed the initial category.
- Sub-categorise the initial category using the mechanistic profilers related to covalent DNA binding. This will eliminate chemicals that contain structural features related to additional covalent mechanisms of action. A robust category should be developed around a single MIE.
- Sub-categorise the category using an appropriate endpoint specific profiler related to mutagenicity. The resulting category has a well-defined mechanistic domain.

Table 6.1 Toxicological data for the seven chemicals assigned to the category developed for the target chemical 3-methylaniline.

Name	Structure	Ames mutagenicity (TA1537)
3-Methylaniline (target chemical)		Read-across prediction: Negative
Aniline		Negative
2-Methylaniline		Negative
4-Methylaniline		Negative
3,4-Dimethylaniline		Negative

Table 6.1 (*Continued*)

Name	Structure	Ames mutagenicity (TA1537)
2,4-Dimethylaniline		Negative
2,5-Dimethylaniline		Equivocal
3,5-Dimethylaniline		Negative

- Sub-categorise using the secondary profilers in order to define the structural domain.

6.2.2 Case Study Two: Category Formation for Skin Sensitisation

This section outlines a case study for building a chemical category to predict the skin sensitisation for the target chemical decanoyl chloride. A number of previous studies have shown that one of the key MIEs for skin sensitisation is covalent protein binding.[3,4] Thus, mechanistic profilers related to this MIE were used in this case study ('protein binding by OECD' and 'protein binding by OASIS'). Skin sensitisation data from the local lymph node assay or the guinea pig maximisation test were utilised as available in the OECD QSAR Toolbox. These data were qualitative giving an indication of skin sensitisation potential only (and not potency).

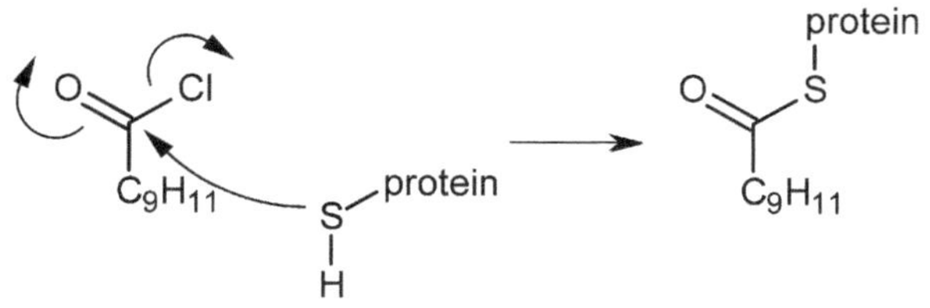

Figure 6.2 Acylation mechanism for decanoyl chloride.

Step 1: Initial profiling using the primary profilers

Profiling decanoyl chloride using the two mechanistic profilers related to covalent protein binding identified an acylation MIE relating to the presence of the acyl chloride moiety (Figure 6.2). This MIE was supported by the profiling results from the endpoint specific 'protein binding for skin sensitisation by OASIS' profiler.

Step 2: Initial category formation and sub-categorisation using the primary profilers

The initial profiling results were then used to retrieve mechanistic analogues from the relevant applicable databases within the OECD QSAR Toolbox. These were:

- Skin sensitisation
- Skin sensitisation ECETOC

Profiling these databases with the 'protein binding by OECD' profiler created an initial category of eight chemicals (including the target chemical). As in the previous case study this initial category required sub-categorisation. The following analysis was carried out:

- Sub-categorisation of the initial category of eight chemicals using the 'protein binding by OECD' profiler (the profiler that was used to develop the initial category). This created a category of seven chemicals.
- Sub-categorisation of the category of seven chemicals using the 'protein binding by OASIS' profiler. This profiler did not identify any additional chemicals to be removed.
- Sub-categorisation of the category of seven chemicals using the 'protein binding for skin sensitisation by OASIS' profiler. As in the case of step 2 this sub-categorisation did not identify any additional analogues that needed removing from the category.

Step 3: Empiric sub-categorisation using the secondary profilers

As in the previous case study it was necessary to define the structural domain of the mechanistic category developed (in this case for skin sensitisation) in Step 2. Thus, the following sub-categorisations were carried out on the category of seven chemicals:

- Sub-categorisation of the category of seven chemicals using the chemical elements profiler. This profiling showed each of the analogues in the category to contain the same elements that were present in the target chemical. Thus, no chemicals were removed by this sub-categorisation.
- Sub-categorisation of the category of seven chemicals using the organic functional group profiler. This analysis resulted in a final category of four chemicals (three analogues and the target chemical).

Step 4: Data-gap filling via read-across

The sub-categorisation carried out using the primary and secondary profilers resulted in a category that had a well-defined mechanistic and structural domain suitable for read-across. Analysis of the toxicological data suggested that decanoyl chloride is likely to be a skin sensitiser if tested *in vivo* (Table 6.2).

The results of the case study showed that the following approach can be considered as a good method for the development of chemical categories for skin sensitisation.

Table 6.2 Toxicological data for the four chemicals assigned to the category developed for the target chemical decanoyl chloride.

Name	Structure	Skinsensitisation
Decanoyl chloride (target chemical)	$H_{19}C_9$–C(=O)–Cl	Read-across prediction: Positive
Nonanoyl chloride	$H_{17}C_8$–C(=O)–Cl	Positive
Hexadecanoyl chloride	$H_{31}C_{15}$–C(=O)–Cl	Positive
Octadecanoyl chloride	$H_{33}C_{16}$–C(=O)–Cl	Positive

Category Formation Case Studies

- Profile the target chemical to identify a potential MIE using mechanistic profilers related to covalent protein binding and an endpoint specific profiler related to skin sensitisation. Confidence is gained in the profiling results if the mechanistic and endpoint specific profilers are in agreement.
- Identify chemicals capable of eliciting the same MIE as the target chemical from suitable toxicological databases. This can be achieved using the mechanistic profilers utilised in Step 1. The resulting category is termed the initial category.
- Sub-categorise the initial category using the mechanistic profilers related to covalent protein binding. This will eliminate chemicals that contain structural features related to additional covalent mechanisms of action. A robust category should be developed around a single MIE.
- Sub-categorise the category using an appropriate endpoint specific profiler related to skin sensitisation. The resulting category has a well-defined mechanistic domain.
- Sub-categorise using the secondary profilers in order to define the structural domain.

6.2.3 Case Study Three: Category Formation for Aquatic Toxicity

The third case study outlines the formation of a chemical category for aquatic toxicity, specifically the 96 hour *Pimephales promelas* assay. Previous research has shown there to be several mechanisms of action for aquatic toxicity, the most common being hydrophobicity dependent narcosis (where hydrophobicity is measured as the logarithm of the octanol:water partition coefficient, log P).[5] In addition, a smaller, but still significant, number of chemicals elicit their toxicity via the formation of a covalent bond with proteins.[6] This case study outlines category formation for chemicals capable of forming such a covalent bond with proteins using cyclohexanecarbaldehyde as the target chemical.

Step 1: Initial profiling using the primary profilers

The profiling results from the two mechanistic profilers ('protein binding by OECD' and 'protein binding by OASIS') for cyclohexanecarbaldehyde suggested the potential for covalent adduct formation via a Schiff base mechanism (Figure 6.3). This was supported by the results from two of the three endpoint specific aquatic toxicity profilers ('aquatic toxicity classification by ECOSAR' and 'acute aquatic toxicity MOA by OASIS') that identified the presence of the aldehyde moiety as a key structural feature associated with aquatic toxicity. The third endpoint specific profiler ('acute aquatic toxicity classification by Verhaar') also showed this chemical to be potentially reactive towards proteins.

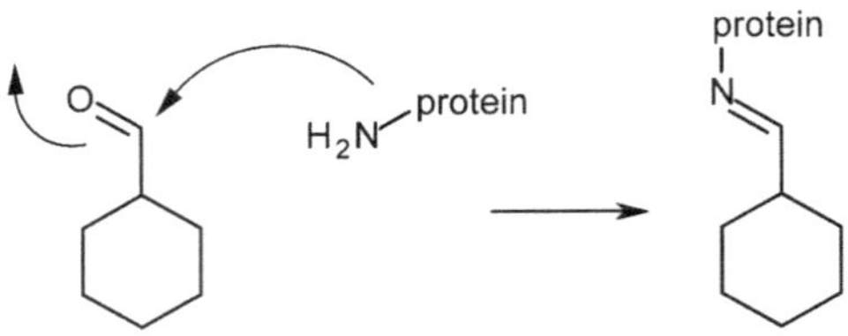

Figure 6.3 Schiff base formation mechanism for cyclohexanecarbaldehyde.

Step 2: Initial category formation and sub-categorisation using the primary profilers

The initial profiling results were then used to retrieve mechanistic analogues from the relevant applicable databases within the OECD QSAR Toolbox. These were:

- Aquatic ECETOC
- Aquatic Japan MoE
- Aquatic OASIS
- Aquatic US-EPA ECOTOX

Searching the four applicable databases resulted in the development of an initial category of 120 chemicals (119 analogues and the target chemical). The following sub-categorisations were then required to ensure the category contained analogues acting via a single mechanism of action:

- Sub-categorisation of the initial category of 120 chemicals using 'protein binding by OECD' profiler (the profiler that was used to develop the initial category). This sub-categorisation resulted in a category of 75 chemicals.
- Sub-categorisation of the category of 75 chemicals using the 'protein binding by OASIS' profiler. This sub-categorisation resulted in a category of 75 chemicals.
- Sub-categorisation of the category of 75 chemicals using the 'aquatic toxicity classification by ECOSAR' profiler. This sub-categorisation resulted in a category of 46 chemicals.
- Sub-categorisation of the category of 46 chemicals using the 'acute aquatic toxicity MOA by OASIS' profiler. This sub-categorisation resulted in a category of 44 chemicals.
- Sub-categorisation of the category of 44 chemicals using the 'acute aquatic toxicity classification by Verhaar' profiler. This sub-categorisation resulted in a category of 35 chemicals (34 analogues and the target chemical).

Step 3: Empiric sub-categorisation using the secondary profilers

As in the previous case studies it was necessary to define the structural domain of the mechanistic category developed in Step 2. Thus, the following sub-categorisations were carried out on the category of 35 chemicals:

- Sub-categorisation of the category of 35 chemicals using the 'organic functional group' profiler. This sub-categorisation resulted in a category of 16 chemicals.
- Sub-categorisation of the category of 16 chemicals using the 'chemical elements' profiler. This sub-categorisation resulted in a category of 16 chemicals (15 analogues and the target chemical).

Step 4: Data-gap filling via read-across

The sub-categorisation performed using the primary and secondary profilers resulted in a category with a well-defined mechanistic and structural domain. Analysis of the toxicological data suggested that cyclohexanecarbaldehyde would have an LC_{50} value equal to 17.5 mgL^{-1} when tested in the 96 hour *Pimephales promelas* assay (Table 6.3).

The results of the case study showed that the following approach can be considered as a good method for the development of chemical categories for acute aquatic toxicity.

- Profile the target chemical using the two mechanistic profilers relevant to covalent protein binding and the three endpoint specific profilers for aquatic toxicity. Confidence in the profiling results is gained if one (or both) of the mechanistic profilers are supported by the results from the endpoint specific profilers.
- Define the initial chemical category by profiling the four databases, relevant to acute aquatic toxicity, using the one of the mechanistic profilers related to covalent protein binding.
- Sub-categorisation using the mechanistic profilers related to covalent protein binding. This sub-categorisation eliminates chemicals that contain structural alerts related to alternative mechanisms of action.
- Sub-categorisation using the endpoint specific profilers relevant to acute aquatic toxicity. These profilers identify known structural alerts related to aquatic toxicological data.
- Sub-categorisation using the secondary profilers in order to define the structural domain. One should use a combination of the empiric profilers (it is recommended to use the 'organic functional group' and 'chemical elements' profilers in the majority of cases) to restrict the structural domain of the category so that it is similar to that of the target chemical.

6.2.4 Case Study Four: Category Formation for Oestrogen Receptor Binding

This case study relates to read-across predictions for oestrogen receptor binding. This is one of many key effects that may be related to endocrine disruption. Historically, there have been many Quantitative Structure-Activity Relationships for the prediction of oestrogen receptor binding,[7–9] but fewer attempts to develop categories for either oestrogen binding or endocrine disruption. There is an

Table 6.3 Read-across prediction made for the *Pimephales promelas* 96 hr LC_{50} endpoint for cyclohexanecarbaldehyde.

Name	Structure	LC_{50} (mgL^{-1})	log P
Cyclohexanecarbaldehyde (target chemical)		Read-across prediction: 17.5	2.10
2-Methylbutanal		9.9	1.23
2-Methylpentanal		18.6	1.73
Hexanal		17.8	1.80
Heptanal		12.0	2.29
2-Ethylhexanal		30.1	2.71

important distinction to make here between category formation and read-across for a receptor mediated effect and those for covalent interactions *e.g.* DNA or protein binding or relative unspecific effects such as narcosis. Covalent interactions can be defined in terms of molecular fragments associated with reactivity; these are an ideal basis for defining a category. Receptor mediated effects require different approaches to grouping and currently in the OECD QSAR Toolbox are represented by 2D properties relating to the presence or absence of molecular features and properties. The development of technologies for 3D receptor mediated effects is an area that clearly requires further development.

This particular case study outlines how to build a category for an alkylphenol, 4-(2,2-dimethylpentyl)phenol, shown in Figure 6.4. Anecdotally alkylphenols are known to bind to the oestrogen receptor[10], although there appear to be no data for this particular chemical. The relative oestrogen receptor binding affinity (ERBA) was predicted by read-across using two profilers, namely those for Estrogen Receptor Binding (a general mechanistic profiler) and the rainbow trout oestrogen receptor binding (rtER) expert system from the US EPA. The ERBA profiler is based on structural and parametric rules. The ERBA classifies chemicals as a binder or a non-binder on the basis of molecular weight and structural characteristics. Specifically for binders, chemicals are classified as very strong, strong, moderate or weak binders on the basis of structural characteristics (*e.g.* rings, hydroxy groups) and molecular weight. The application of the ERBA binder has been evaluated by Mombelli.[11] The rtER Expert System is based on a number of structural rules and physico-chemical properties that can not only provide predictions, but also allow for grouping. It is the grouping aspects that have been utilised in this case study. Both profilers are strongly linked to the AOP for oestrogen receptor binding as published by Ankley *et al.*[12]

The two profilers will be considered separately in this case study, this is different to case studies 1–3 where information from different profilers was used to help sub-categorise the grouping. The reason for this is that the rtER Expert System is a predictive tool that allows for category formation on the basis of a decision (incorporating physico-chemical properties) rather than a solely mechanistic profiler. Both profilers were supplemented with information from the "organic functional group" empiric profiler.

Step 1: Initial profiling using the primary (ERBA) profiler

4-(2,2-dimethylpentyl)phenol was profiled using the ERBA (general mechanistic) profiler. As would be expected, the target molecule was associated with binding to the oestrogen receptor and, specifically, was profiled as being a moderate binder due to the hydroxy group on the ring. This profile is associated with compounds of molecular weight between 170 and 200 Da and a non-impaired –OH group (*i.e.* a primary hydroxy group) attached to a 5 or 6-membered carbon ring. Such compounds are commonly associated with binding into site A of the oestrogen receptor binding pocket.[13] In addition, functional groups were identified with the organic functional groups profiler.

Step 2: Initial category formation and sub-categorisation using the primary (ERBA) profiler

The ERBA profiler categorised 4-(2,2-dimethylpentyl)phenol as a moderate binder due to the hydroxy group on the ring. Information was sought for compounds in the same category from the OECD QSAR Toolbox. The relevant databases searched were:

- Estrogen Receptor Binding Affinity OASIS
- Yeast estrogen assay database, University of Tennessee-Knoxville (USA)

Profiling these databases with the 'Estrogen Receptor Binding Affinity' profiler created an initial category of 53 chemicals (52 analogues plus the target chemical). In theory this category could be utilised as it is because it is well-defined and all compounds would be expected to act by the same mechanism.

Step 3: Empiric sub-categorisation using the secondary profilers

Visual inspection of the compounds in the category suggested that further sub-categorisation is required due to the presence of halogen atoms and other rings on molecules. Thus, the following sub-categorisation was undertaken to ensure the structural homogeneity of the category:

- Sub-categorisation of the category of 53 chemicals using the organic functional groups profiler. This showed four functional groups as being present in the target molecule, namely aryl (ring), benzyl, phenol and precursors of quinoid compounds. There is obvious overlap between all function groups and compounds for inclusion in the final category were selected on the basis of containing all four functional groups. This sub-categorisation resulted in a category of ten chemicals (nine analogues and the target). The sub-category appears robust both from a mechanistic and a structural point of view and no further sub-categorisation was undertaken.

Step 4: Data-gap filling via read-across

The category formed from the ERBA and organic functional group profilers revealed ten compounds, including the target. All analogues had oestrogen receptor binding data. Table 6.4 shows that structure of the target chemical and the five closest analogues according to the analysis with log P, also shown are the ER binding assay results. Read-across using these five data gives a relative oestrogen binding affinity of 0.0087%.

The OECD QSAR Toolbox contains an additional profiler suitable for developing chemical categories for toxicity mediated by oestrogen receptor binding. This is the rtER expert system profiler and was used as follows to develop a second category:

Step 1: Initial profiling using the primary (rtER Expert System) profiler

4-(2,2-dimethylpentyl)phenol was also profiled using the rtER (endpoint specific) profiler. The compound is profiled as being an "alkylphenol". The rules to establish this are that it contains a carbon atom, a ring with a phenol

Table 6.4 Structures, oestrogen receptor binding affinity and log P of 4-(2,2-dimethylpentyl)phenol and its five closest analogues identified by the ERBA and rtER profilers.

Name	Structure	log P	Oestrogen Receptor Binding Affinity (%)
4-(2,2-dimethylpentyl)phenol (target chemical)		4.90	Read-across prediction: 0.0087
4-phenethylphenol (ERBA closest analogue)		4.26	0.0020
4-tert-hexylphenol (ERBA and rtER closest analogue)		4.40	0.010

Table 6.4 (*Continued*)

Name	Structure	log P	Oestrogen Receptor Binding Affinity (%)
4-hexylphenol (ERBA and rtER closest analogue)		4.52	0.0067
4-tert-heptylphenol (ERBA and rtER closest analogue)		4.90	0.014
4-heptylphenol (ERBA and rtER closest analogue)		5.01	0.011
4-octylphenol (rtER closest analogue)		5.50	0.0047

fragment and an alkyl substituent, also log P is greater than 1.3, and it does not infringe on any special rules. A training set (with measured rainbow trout oestrogen receptor binding data) is provided in the Toolbox, this comprises 21 alkylphenols, the majority of which could be considered direct analogues of the target compound. In addition, functional groups were identified with the organic functional groups profiler.

Step 2: Initial category formation and sub-categorisation using the primary (rtER) profiler

The rtER profiler categorised 4-(2,2-dimethylpentyl)phenol as an alkylphenol. Information was sought for compounds in the same category from the OECD QSAR Toolbox. The relevant databases searched were:

- Estrogen Receptor Binding Affinity OASIS
- Yeast oestrogen assay database, University of Tennessee-Knoxville (USA)

Profiling these databases with the rtER profiler created an initial category of 36 chemicals (35 analogues plus the target chemical). Again, this should be a very well-defined category of chemicals acting by the same mechanism of action. Thus, in theory, it would be possible to use this category as it is.

Step 3: Empiric sub-categorisation using the secondary profilers

Visual investigation of the compounds in the group indicated that this is a very well defined category; specifically all compounds contain a phenolic group and an alkyl substituent. In order to thoroughly ensure the structural homogeneity of the category, the following sub-categorisation was undertaken:

- Sub-categorisation of the category of 36 chemicals using the organic functional groups profiler. This showed four functional groups as being present in the target molecule, namely aryl (ring), benzyl, phenol and precursors of quinoid compounds. This sub-categorisation resulted in a category of 22 chemicals (21 analogues and the target). The sub-category appeared robust both from a mechanistic and structural point of view and no further sub-categorisation was undertaken.

Step 4: Data-gap filling via read-across

The category formed from the rtER and organic function group profilers provided a category of 22 compounds, including the target. All analogues had oestrogen receptor binding data. Table 6.4 shows the structure of the target chemical and the five closest analogues according to the analysis with log P; also shown are the ER binding assay results. It is reassuring, although of no surprise, that four of the five closest analogues are common between the two profilers. Naturally, therefore, the read-across using the five data from the rtER profiler gives a relative oestrogen binding affinity, very similar to that for the ERBA profiler, of 0.0093%.

The following outline can be considered a good general approach for the development of chemical categories for oestrogen receptor binding.

- Profile the target chemical with either, or preferably both, of the profilers ERBA and rtER.
- If the compound is classified as a potential oestrogen receptor binder, analyse the categories obtained. If the category is robust from a structural perspective then no more sub-categorisation may be required — in this case proceed directly to read-across. If the compound is classified as a non-binder, then go to Step 4.
- For compounds that may bind to the oestrogen receptor, sub-categorisation may be required on the basis of organic functional groups to ensure a structurally robust category.
- For compounds categorised as non-binders, further analysis may or may not be warranted. Grouping may be required on the basis of structural similarity and/or organic functional groups to elicit read-across. However, read-across may be difficult in terms of finding suitable data as the most testing has been directed to those structures likely to have, or be similar to those that have, the capability of binding to the oestrogen receptor.

6.2.5 Case Study Five: Category Formation for Repeated Dose Toxicity

This case study describes some of the excellent work of Sakuratani and co-workers from the National Institute of Technology and Evaluation (NITE), Tokyo, Japan and the application of the Hazard Evaluation Support System (HESS) for making these predictions. The HESS system is now included as part of the OECD QSAR Toolbox, or is available as a stand-alone download (http://www.safe.nite.go.jp/english/kasinn/qsar/hess-e.html). Both systems are currently available free of charge, although it is suggested that the user may wish to use the version in the OECD QSAR Toolbox as it provides more comprehensive databases and access to other profilers.

This particular case study on the repeated dose toxicity of nitrobenzenes describes and summarises the work published by Sakuratani et al.[14] Reference is also made to the findings and methodology of Sakuratani et al.[15] on repeated dose toxicity of anilines and Yamada et al.[16] with regard to predicting the repeated-dose hepatotoxicity of allyl esters. The work described in this case study, presented first by Sakuratani et al.,[17] utilises the HESS tool.[14,18]

The aim of the original study by Sakuratani et al.[17] was to demonstrate how a category could be developed for nitrobenzenes to allow for read-across of repeated dose toxicity. Specifically, nitrobenzenes are known to produce toxic responses through the action of their metabolites which are anilines. This knowledge is sufficient to form the basis of a mechanistically derived category, noting also that Sakuratani et al.[15] derived a separate category for anilines.

It should be noted that there is a key difference in this case study as compared to the previous case studies. The previous case studies used profilers to group chemicals and derive categories. The example provided by Sakuratani

Category Formation Case Studies

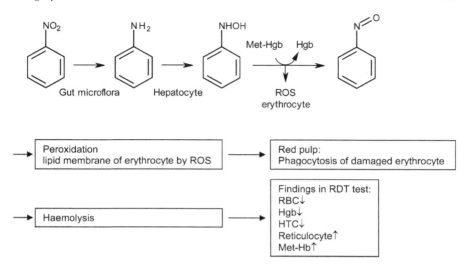

Figure 6.4 Summary of the Adverse Outcome Pathway for haemolytic anaemia induced by the nitrobenzenes.

et al.[17] assumes prior knowledge of a possible grouping *i.e.* the nitrobenzenes are linked by common metabolites. This provides an excellent starting place for grouping. Evidence is provided to support this mechanistic basis with reference to a simplified Adverse Outcome Pathway (AOP) for haemolytic anaemia induced by the nitrobenzenes (following metabolism to the aniline). The AOP is summarised in Figure 6.4 and is supported by evidence from the literature.[19–22]

This case study can be repeated using the profilers and data available in the HESS system, which are available in the OECD QSAR Toolbox. The HESS system (termed "repeated dose (HESS)") in the Toolbox (ver 3.1) comprises 33 pre-defined categories that are known to be related to repeated dose oral toxicity. These comprised structural classes for common industrial chemicals. There are three profilers for nitrobenzenes in the HESS system, being related to different effects: haemolytic anaemia with methemoglobinemia; heptatoxicity; and testicular toxicity. The HESS system ranks the profilers in terms of reliability from A (most well-known, supported by experimental evidence and hence reliable) to C (least well-known). For the nitrobenzenes, haemolytic anaemia is ranked A, whereas heptatoxicity and testicular toxicity are ranked C. Therefore, the profiler for haemolytic anaemia was utilised, not least as it is supported by the AOP for this effect.

Steps 1–3: Initial profiling, category formation and sub-categorisation.

The first three steps of the categorisation process will be described together for the example of the repeated dose toxicity of nitrobenzenes. The information given here summarises that provided by Sakuratani *et al.*[17] The initial profiling is implicit in the definition of the category, *i.e.* that these

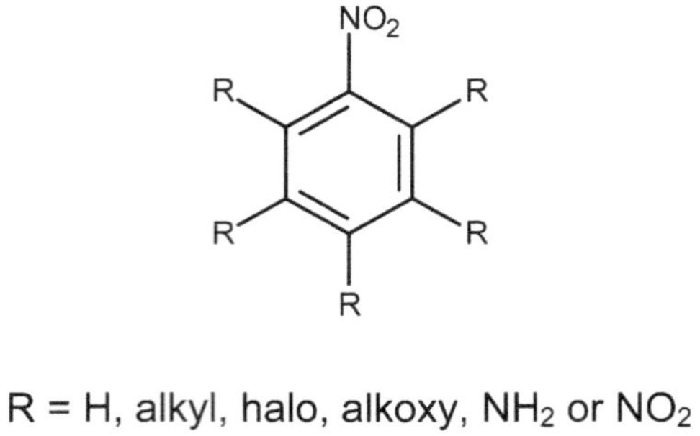

R = H, alkyl, halo, alkoxy, NH₂ or NO₂

Figure 6.5 Structural boundaries of nitrobenzene category for haemolytic anaemia induced by the AOP shown in Figure 6.4.

compounds are nitrobenzenes. According to Sakuratani et al.[17] and the profiler in the HESS system the structural boundaries for this category (defined with reference to the AOP) are shown in Figure 6.5.

Searching of the HESS and other databases for this specific category identified 25 repeated dose toxicity reports for 24 different nitrobenzenes. The compounds and results are shown in Table 6.5. For the results for the Lowest Observed Effect Level (LOEL), not all are shown in Table 6.5 but are available in more detail in Sakuratani et al.[17] Thorough investigation of the LOEL values does confirm that it is liver effects and haemolytic anaemia in particular that have the lowest LOEL of all organ level toxicities (other values are not shown in Table 6.5). For many of the test results, there is further evidence that for the induction of haemolytic anaemia from the reported production of Met-Hgb and the binding of Hgb.[20-22] This, along with the evidence provided by the AOP, provides robust evidence in support of this category.

The category was further assessed in terms of sub-categorisation based on biological data and the test results. Such an approach is not standard, but helps to build a picture of the category and its properties which will be useful for read-across. Thus, for 1-chloro-2,4-dinitrobenzene, haemolytic anaemia was not observed; however, investigation of the original data showed that the maximum administration dose was low (0.15 mmol/kg per day) and it was unclear if haemolytic anaemia could be induced at higher doses. Conversely, nitrobenzenes substituted with a hydroxyl or acid group (the twelve latter compounds in Table 6.5) were seldom associated with haemolytic anaemia at low doses in repeated dose tests. This is explained with regard to their polarity and thus low hydrophobicity which prevents them from being distributed significantly into hepatocytes or erythrocytes. Thus, the nitrobenzene category defined by Figure 6.5 can be further defined in terms of hydrophobicity.

Step 4: Data-gap filling via read-across

Sakuratani et al.[17] state that within the category defined by Figure 6.5, nitrobenzenes assessed under similar repeated dose test conditions can be assumed to be capable of haemolytic anaemia. Thus, within the boundaries of

Table 6.5 Chemicals assigned to the nitrobenzenes category, summary of the repeated dose test protocol and Lowest Observed Effect Level (LOEL) for haemolytic anaemia and liver effects.

Name	Test	Administration Route	Period (days)	Haemolytic anaemia LOEL (mmol/kg per day)	Liver effect LOEL (mmol/kg per day)	Reference
Nitrobenzene	OECD407	Gavage	28	0.041	0.041	23
Nitrobenzene	OECD422	Gavage	42	0.16	0.16	24
2-Nitrotoluene	NTP	Feed	90	0.33	0.33	25
3-Nitrotoluene	NTP	Feed	90	1.2	4.8	25
4-Nitrotoluene	NTP	Feed	90	0.31	0.31	25
1,2-Dichloro-3-nitrobenzene	OECD422	Gavage	42	0.13	0.13	26
2,4-Dichloronitrobenzene	OECD422	Gavage	42	0.042	0.042	26
1,2-Dichloro-4-nitrobenzene	OECD422	Gavage	42	0.1	0.52	27
2-Nitroanisole	NTP	Feed	90	0.065	0.065	25
1-Chloro-2,4-dinitrobenzene	OECD422	Gavage	42	–	–	26
3-Nitrobenzenamine	OECD407	Gavage	28	0.11	0.36	26
4-Chloro-2-nitroaniline	OECD422	Gavage	42	0.35	1.7	27
4-Nitro-2-anisidine	OECD407	Gavage	28	1.8	1.8	26
N,O-Di(2-hydroxyethyl)-2-amino-5-nitrophenol	NTP	Feed	90	–	1.3	25
4-Nitrophenol sodium salt	OECD407	Gavage	28	–	–	26
2-Amino-4-nitrophenol	NTP	Gavage	90	–	3.2	25
2-Amino-5-nitrophenol	NTP	Gavage	90 days	–	1.3	25
2,4-Dinitrophenol	OECD407	Gavage	28 days	–	0.43	26
2-sec-Butyl-4,6-dinitrophenol	OECD422	Gavage	42 days	0.01	–	26
2,4,6-Trinitrophenol	OECD407	Gavage	28 days	0.44	0.44	26
Sodium 3-nitrobenzenesulfonate	OECD407	Gavage	28 days	–	–	26
2-Methyl-5-nitrobenzenesulfonic acid	OECD422	Gavage	42 days	–	3.2	26

Table 6.5 (Continued)

Name	Test	Administration		Haemolytic anaemia LOEL (mmol/kg per day)	Liver effect LOEL (mmol/kg per day)	Reference
		Route	Period (days)			
3-Nitrobenzoic acid	OECD422	Gavage	42 days	3	3	27
4-Nitrobenzoic acid	NTP	Feed	90 days	0.24	0.24	25
4-Hydroxy-3-nitrobenzenearsonic acid	NTP	Feed	90 days	–	0.015	25

this category, read-across can be performed, in this case it is the mean and 95% confidence interval of the LOEL for haemolytic anaemia for the compounds (the first thirteen compounds listed in Table 6.5), this value being 0.39 ±0.35 mmol/kg per day. Further assessment of these compounds enabled definition of this category to be developed for nitrobenzenes as denoted by Figure 6.5. Chemicals within this category were found to have log P values between 1.8–3.1 and molecular weights between 123 and 203 gmol^{-1}.

The results of the case study showed that the following approach can be considered a good method for the use of chemical categories for repeated dose toxicity.

- Profile the target chemical using the tools such as the HESS repeated dose profiler. Should this reveal that the target chemical belongs to a category then this should give greater confidence.
- If the target cannot be profiled using the HESS Profiler: use the functional group profilers available in the OECD QSAR Toolbox to categorise the chemical. Sub-categorisation may be required; this may need to be performed in relation to the available toxicity data.
- Search the available toxicity data for repeated dose effects. At this point expert analysis will be required to interpret the LO(A)EL values to determine if there is consistency across organ level effects. If such consistency can be found, does it relate to a recognisable mechanism or mode of action or, preferably, an AOP. The data in the category should be investigated for structural and physico-chemical trends to support the read-across and category definition.

Some issues specific to predicting repeated dose toxicity through read-across must be kept in mind.

- Repeated dose effects are some of the most complex to predict from a category and read-across approach. Care and expertise are required at all stages of the analysis. The user must be aware that attempting to predict a LO(A)EL value is difficult. The success summarised above and reported by Sakuratani et al.[17] is due to the strength and mechanistic robustness of the category.

- The HESS tool provides an ideal starting point for category formation. It is hoped that the categories available within it will be expanded and further molecular initiating events identified. If a target compound falls outside of these profilers, others such as functional group and similarity may be used with caution.

6.3 Similarity-based Case Studies

The above section outlined the use of mechanism based profilers to group chemicals into categories suitable for read-across. These methods work well if sufficient knowledge exists of potential MIEs and this information has been encoded into structural alerts. However, for a number of endpoints such information regarding potential MIEs does not exist or remains to be

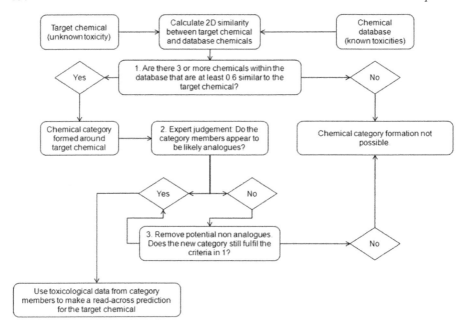

Figure 6.6 Flow chart for using 2D chemical similarity to form a category reproduced from Enoch et al.[24]

elucidated. In these cases alternative chemoinformatic-based methods using 2D similarity methods can be useful for grouping chemicals. The flow chart shows (Figure 6.6) how 2D similarity can be used to build a chemical category. Such analysis can be carried out using tools such as Toxmatch.[23]

6.3.1 Case Study Six: Category Formation for Teratogenicity

A recent study outlined the use of chemical similarity to develop chemical categories for teratogenicity using data from the US FDA.[24] The study involved the development of categories for a series of target chemicals by identifying similar chemicals from a teratogenicity database. The database contained chemicals that had been previously classified into one of five teratogenicity classes as a result of analysis by the US FDA (A, B, C, D and X, where D and X are of significant concern).[25] The category formation analysis was carried out using the 2D-similarity metrics in the freely available Toxmatch software[23], (see Chapter 4). The results showed that this type of analysis produced robust categories in which chemicals identified as being similar to the target chemical, and thus part of the category, were in the same teratogenicity class. For example, three chemicals were identified as being sufficiently similar to ethynodiol diacetate (Figure 6.7). These chemicals were then used to make a read-across prediction that ethynodiol diacetate should be assigned to teratogenicity class D using a weight of evidence approach (or class

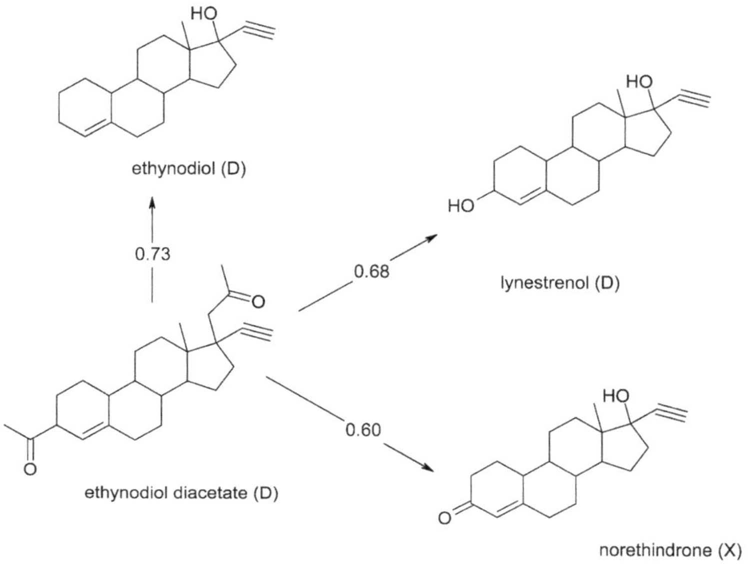

Category Formation Case Studies 151

Figure 6.7 Chemical category developed using Toxmatch for the target chemical ethynodiol diacetate allowing for the prediction of teratogenicity (similarity measures as shown on the arrows, US FDA teratogenicity class shown in parenthesis).

X taking a worst case scenario). The authors of the study highlighted that inspection of the actual FDA classification for ethynodiol diacetate showed it to have been previously assigned to class D.

The results of the case study showed that the following approach can be considered a good method for the use of chemical similarity for the formation of chemical categories for endpoints for which detailed knowledge of potential MIEs is currently limited:

- Identify an endpoint for which no information regarding potential MIEs exists and for which a data-gap exists for a target chemical.
- Encode the target chemical and the chemicals in a suitable database containing toxicological data as suitable bit-strings (see Chapter 2). This can be achieved using freely available software such as Toxmatch.
- Measure the similarity between the target chemical and the chemicals in the database using a suitable measurement of chemical similarity and cut-off (the Tanimoto coefficient coupled with a cut-off of 0.6 has been shown to be useful in forming robust categories).
- Visually inspect the chemicals identified as being 'similar' in step 3. Remove any obviously dissimilar chemicals from the category.

6.4 Conclusions

This chapter has outlined category formation case studies for a number of different toxicological endpoints. It has highlighted that one of the most important methods for category formation is based around mechanistic knowledge of the molecular initiating event. The examples outlined show that the mechanistic knowledge encoded as profilers leads to robust and transparent categories within which a read-across prediction can be made. Finally, the chapter has also demonstrated that for toxicological endpoints for which mechanistic information is lacking chemical similarity can be used to form categories.

Acknowledgements

Funding from the European Community's Seventh Framework Program (FP7/2007–2013) COSMOS Project under grant agreement no. 266835 and from Cosmetics Europe is gratefully acknowledged. Funding from the European Chemicals Agency (Service Contract No. ECHA/2008/20/ECA.203) is also gratefully acknowledged.

References

1. The OECD QSAR Toolbox. (available from: http://www.qsartoolbox.org/)
2. B. N. Ames, E. G. Gurney, J. A. Miller and H. Bartsch, Carcinogens as frameshift mutagens: Metabolites and derivatives of 2-acetylaminofluorene and other aromatic amine carcinogens. *Proc. Natl. Acad. Sci. U. S. A*, 1972, **69**, 3128.
3. A. O. Aptula, G. Patlewicz and D. W. Roberts, Skin sensitization: Reaction mechanistic applicability domains for structure-activity relationships. *Chem. Res. Toxicol.*, 2005, **18**, 1420.
4. D. W. Roberts, G. Patlewicz, P. S. Kern, F. Gerberick, I. Kimber, R. J. Dearman, C. A. Ryan, D. A. Basketter and A. O. Aptula, Mechanistic applicability domain classification of a local lymph node assay dataset for skin sensitization. *Chem. Res. Toxicol.*, 2007, **20**, 1019.
5. H. J. M. Verhaar, C. J. van Leeuwen and J. L. M Hermens, Classifying environmental pollutants. 1: Structure-activity relationships for prediction of aquatic toxicity, *Chemosphere*, 1992, **25**, 471.
6. S. J. Enoch, C. M. Ellison, T. W. Schultz and M. T. D Cronin, A review of the electrophilic reaction chemistry involved in covalent protein binding relevant to toxicity. *Crit. Rev. Toxicol.*, 2011, **41**, 783.
7. I. Tsakovska, I. Pajeva, A. Petko and A. Worth, Recent advances in the molecular modeling of estrogen receptor-mediated toxicity, in *Advances in Protein Chemistry and Structural Biology*, ed. C. Christov, Academic Press, Burlington, 2011, p. 217.

8. M. Novic and M. Vracko, QSAR models for reproductive toxicity and endocrine disruption activity. *Molecules*, 2010, **15**, 1987.
9. X. Xu, W. Yang, Y. Li and Y. Wang, Discovery of estrogen receptor modulators: a review of virtual screening and SAR efforts. *Expert Opin. Drug Discov.*, 2010, **5**, 21.
10. M. R. Servos, Review of the aquatic toxicity, estrogenic responses and bioaccumulation of alkylphenols and alkylphenol polyethoxylates. *Water Qual. Res. J. Can.*, 1999, **34**, 123.
11. E. Mombelli, Evaluation of the OECD (Q)SAR Application Toolbox for the profiling of estrogen receptor binding affinities. *SAR QSAR Environ. Res.*, 2012, **23**, 37.
12. G. T. Ankley, R. S. Bennett, R. J. Erickson, D. J. Hoff, M. W. Hornung, R. D. Johnson, D. R. Mount, J. W. Nichols, C. L. Russom, P. K. Schmieder, J. A. Serrano, J. E. Tietge and D. L. Villeneuve, Adverse outcome pathways: A conceptual framework to support ecotoxicology research and risk assessment. *Environ. Toxicol. Chem.*, 2010, **29**, 730.
13. P. K. Schmieder, G. T. Ankley, O. Mekenyan, J. D. Walker and S. Bradley, Quantitative structure-activity relationship models for prediction of estrogen receptor binding affinity of structurally diverse chemicals. *Environ. Toxicol. Chem.*, 2003, **22**, 1844.
14. Y. Sakuratani, H. Q. Zhang, S. Nishikawa, K. Yamazaki, T. Yamada, J. Yamada, K. Gerova, G. Chankov, O. Mekenyan and M. Hayashi, Hazard Evaluation Support System (HESS) for predicting repeated dose toxicity using toxicological categories. *SAR QSAR Environ. Res.*, **24**, 617.
15. Y. Sakuratani, S. Sato, S. Nishikawa, J. Yamada, A. Maekawa and M. Hayashi, Category analysis of the substituted anilines studied in a 28-day repeat-dose toxicity test conducted on rats: Correlation between toxicity and chemical structure. *SAR QSAR Environ. Res.*, 2008, **19**, 681.
16. T. Yamada, Y. Tanaka, R. Hasegawa, Y. Sakuratani, J. Yamada, E. Kamata, A. Ono, A. Hirose, Y. Yamazoe, O. Mekenyan and M. Hayashi, A category approach to predicting the repeated-dose hepatotoxicity of allyl esters. *Reg. Toxicol. Pharmacol.*, 2013, **65**, 189.
17. Y. Sakuratani, H. Q. Zhang, S. Nishikawa, K. Yamazaki, T. Yamada, J. Yamada and M. Hayashi, Categorization of nitrobenzenes for repeated dose toxicity based on adverse outcome pathways. *SAR QSAR Environ. Res.*, 2013, **24**, 35.
18. M. Hayashi and Y. Sakuratani, Development of an evaluation support system for estimating repeated dose toxicity of chemicals based on chemical structure, in *New Horizons in Toxicity Prediction*, ed. A. G. E. Wilson, Royal Society of Chemistry, Cambridge, UK, 2011, p. 26.
19. V. Facchini and L. A. Griffiths, The involvement of the gastro-intestinal microflora in nitrocompound-induced methaemoglobinaemia in rats and its relationship to nitro group reduction. *Biochem. Pharmacol.*, 1981, **30**, 931.

20. G. Sabbioni, Hemoglobin binding of nitroarenes and quantitative structure-activity relationships. *Chem. Res. Toxicol.*, 1994, **7**, 267.
21. T. P. Bradshaw, D. C. McMillan, R. K. Crouch and D. J. Jollow, Identification of free radicalsproduced in rat erythrocytes exposed to hemolytic concentrations of phenylhydroxylamine. *Free Rad. Biol. Med.*, 1995, **18**, 279.
22. M. Kiese, Methemoglobinemia: A Comprehensive Treatise. CRC, Cleveland, OH, 1974.
23. Toxmatch software. (*available from* http://ihcp.jrc.ec.europa.eu/).
24. S. J. Enoch, M. T. D. Cronin, J. C. Madden and M. Hewitt, Formation of structural categories to allow for read-across for teratogenicity. *QSAR Combi. Sci.*, 2009, **28**, 696.
25. G. C. Briggs, R. K. Freeman and S. J. Yaffe, *Drugs in Pregnancy and Lactation*. 6th ed., Lippincott Williams and Wilkins, Philadelphia, 2002.

CHAPTER 7

Evaluation of Categories and Read-Across for Toxicity Prediction Allowing for Regulatory Acceptance

M. T. D. CRONIN

School of Pharmacy and Chemistry, Liverpool John Moores University, Byrom Street, Liverpool L3 3AF, England
E-mail: m.t.cronin@ljmu.ac.uk

7.1 Introduction

The successful use of predictive methods requires assessment of the validity of the model, in this case the category formation and read-across strategies, and whether the prediction for a particular compound is valid. This two stage process requires a framework to evaluate the model and the prediction. This process is required to allow for consideration of predictions for possible regulatory use, for instance, as part of REACH dossiers. The prediction must also be documented adequately to ensure that appropriate and sufficient evidence is presented to those who may review the prediction.

To understand the evaluation process, the key stages to obtaining a valid read-across prediction for toxicity, or any other endpoint, must be understood. Broadly speaking, these are:

- The development of a robust and justifiable category that is relevant to the endpoint being predicted.

- Confidence that the target chemical belongs to the category *i.e.* is within the domain of the category.
- The possibility of performing read-across *i.e.* there is sufficient information to support the prediction.

In order to implement these criteria, practical solutions are required that are achievable, not only for the modeller and user of the model, but also of value to those utilising the prediction both from industry and regulatory agencies. When the practitioner comes to perform their modelling, at the forefront of their mind should be the successful execution of the model and its evaluation. This should be part of the planning process and considered both before the modelling is undertaken and during the prediction process. It is a mistake to not make this an integral part of the prediction and to add it on at the end. Some of the issues that the practitioner should consider are:

- What confidence is there in the category, is it relevant to the target chemical and data? Obviously confidence should be as high as possible and the category and its members need to be designed and selected to achieve this.
- How can the level of confidence in the category, data and read-across prediction be demonstrated and proven? The design of the category should ensure that documentation can be provided and there is sufficient evidence, *e.g.* a mechanistic basis, to support it. This means that appropriate tools must be utilised and the whole modelling and prediction process recorded and documented.
- Determination that any prediction from read-across is demonstrably valid.

At this point it must become obvious to the reader and practitioner that each step is subjective and requires expert use and opinion. The information below, in addition to that provided in the guidance documents (see Table 1.4) and the excellent overview from Patlewicz *et al.*[1] is intended to provide practitioners instruction on how to assign confidence and determine the validity of a prediction. Please note that the purpose here is not to "validate" a category or QSAR. This chapter aims to outline the process of "good modelling" to allow for the evaluation of a prediction and, where possible, regulatory acceptance.

7.2 Assigning Confidence to the Robustness of a Category

One of the first tasks that must be undertaken to evaluate a read-across prediction is to determine the robustness of the category. Robustness implies that the category is transparent, interpretable and defensible. As defined in Chapter 2, a category is a group of "similar" chemicals. Thus it is logical to assume a robust category will comprise chemicals that are demonstrably similar. Therefore, the strength and security afforded by the analogue approach should become immediately obvious, *i.e.* a group of compounds with strong structural similarity or varying in only the length of a carbon chain. Furthermore, groupings of compounds made using structural similarity algorithms will, within reason, be robust — although experience shows that care must be taken in using these approaches. It will be more difficult to assign confidence to categories based on mechanistic or endpoint profilers. This is because the compounds in the category may not necessarily be structural analogues. At this point, the category will need to be justified with evidence that all compounds have the capability to act by the same mechanism. Tools such as the OECD QSAR Toolbox (see Section 4.3) provide such evidence, linking chemistry to toxicology. If other approaches are undertaken, then an equivalent level of detail and knowledge will be required.

Given that a group of similar chemicals has been selected, the second criterion that will determine confidence will be the number of compounds in the category. This may seem a trivial point to address but there are a number of issues that need to be considered:

- Grouping tools, such as the OECD QSAR Toolbox, select chemical structures for grouping on the basis that they are similar, rather than identifying those with toxicity data that may be useful for read-across. Thus an initial group or category formed in the Toolbox may contain hundreds, if not thousands, of structures. This does not necessarily indicate that this is a robust group, rather that the boundaries (or definition) of the cluster maybe too broad and hence sub-categorisation may be required.
- The presence of compounds or structures in a category does not necessarily imply the availability of toxicity data from which to obtain a read-across prediction. Hence, even the largest category (see previous point) may be unusable for prediction if there are no or too few data, or data of insufficient quality.
- Depending on the use of the prediction, more evidence may be required for the regulatory acceptance of a negative prediction. This implies that there must be strong evidence of structural or mechanistic similarity and a "significant" number of compounds in the category with negative results.

Therefore there is always likely to be a balance between the size of the group or category, the degree of similarity of structures with regard to the endpoint being modelled and the number of available data. No hard and fast rules can be developed for this, indeed it may be entirely feasible to perform a read-across from a single compound *e.g.* for a positive read-across prediction for a close structural and mechanistic analogue for an endpoint such as mutagenicity. Thus, the subjectivity of this approach must be appreciated.

7.3 Assigning Confidence to the Read-Across Prediction

There are no firm or well defined rules for assessing the quality of a read-across prediction and/or assigning confidence to it. Indeed, a rigid framework to assess read-across predictions is not desirable as it may be too restrictive.[1,2] There is, however, good guidance and comment provided by Patlewicz *et al.*[1]

and in more detail by ECETOC.[3] Forthcoming case studies, such as those proposed within the Read-Across Assessment Framework (RAAF) from the European Chemicals Agency (ECHA) will assist in providing knowledge and insight on how to provide evidence to substantiate read-across predictions.[4-6]

High confidence will be assigned to a read-across prediction when there is strong proof it is valid. Therefore, assuming the category is robust and membership is assured, the read-across will be reliant on the similarity of the target chemical to the other chemicals in the category and the quality and number of data. Some guidance suggests that data for at least ten compounds must be available for a category to be significant.[4] However, this may be an unrealistic goal and pragmatism will dictate how many data will be used *i.e.* in most cases it would be better to have fewer data for a structurally and/or mechanistically similar group of compounds than extending the category with less confidence in category membership, simply in an attempt to increase numbers. It is also inevitable that, at least initially, higher confidence will be given to structural analogues rather than mechanistically developed categories.

For a well developed read-across prediction it would be expected that the toxicity data are consistent. Thus for a categoric toxicity endpoint (*i.e.* toxic or non-toxic, mutagenicity being an example) all compounds in the category have the same result *i.e.* either toxic or non-toxic. Of course, the category approach implicitly allows for contradictory data, but there will need to be a rational explanation for these inconsistencies. Examples of reasons why similar compounds may demonstrate different activities include metabolism, structural mitigating factors (*e.g.* steric hindrance around a reactive functional group), solubility, bioavailability or impurities (see below). For continuous toxicity data (*i.e.* potency values such as acute LD_{50}), predictable trends would be expected with descriptors relevant to activity. Some relationships between potency and molecular properties or descriptors can, of course, form the basis of local QSARs. Thus, consistency in the data within a category, accepting rationally explained exceptions will raise the confidence associated with a read-across prediction.

A further aspect that must be addressed to assess the confidence that may be associated with a read-across prediction is that of impurities. ECETOC provide good guidance here.[3] For instance, impurities $> 0.1\%$ must be identified for CMR, PBT and R50/R53 (very toxic to aquatic organisms; may cause long-term adverse effects in the aquatic environment) endpoints and $> 1\%$ for other endpoints. The impurities must be identified and characterised fully. One possibility is that these will be structurally similar to the target chemical (thus read-across could itself be applied to the impurity), but QSARs (or other read-across) may be required to assess the potential hazard of impurities.

7.3.1 Weight of Evidence to Support a Prediction

It is inevitable that weight of evidence (WoE) from all sources of information will help support and provide evidence for a read-across prediction. There is no defined framework for using WoE to support a prediction, although much useful guidance is provided by Patlewicz et al.[1] and ECETOC.[3] Thus, to support a read-across prediction, evidence can be brought in not only from the category but also other sources. For instance, for a read-across prediction of acute fish toxicity for a given species, further evidence could be provided from non-target species (i.e. other fish species) and also extrapolation from species from different taxa.[7] Likewise, evidence that a compound may be acting as a non-polar narcotic could be drawn from the lack of protein reactivity (assessed either in silico or in chemico).[8] Other examples include using evidence from "related" endpoints. For instance, skin sensitisation and mutagenicity may be brought about by the same covalent interactions, thus (in certain circumstances) a positive mutagenicity assay may be indicative evidence of skin sensitisation.[9] One key area which should become more important to support WoE is the use of alternative non-test data to support a prediction. The use of in vitro test assay results is well established, but the future brings with it the possibility of routinely using molecular biology data to support category formation, membership and read-across.[10] An excellent example of how data from a Metabolomics database (MetaMap®Tox) can inform grouping for toxicity prediction is provided by van Ravenzwaay et al.[11] Section 4.2 provides more information on use of high throughput screening and -omics data to support category formation.

7.4 Reporting of Predictions

In order to gain acceptance for a prediction of toxicity or a physico-chemical property, the method of prediction and the prediction itself must be properly documented. This implies that:

- the method(s) by which the prediction is made are adequately described;
- the method(s) and data considered are available and transparent;
- the prediction could be repeated (if required);
- the prediction is adequately justified.

The concept of documentation of a QSAR is well established — the so-called QSAR Model Reporting Format (QMRF) was developed based around the OECD Principles for the Validation of (Q)SARs.[12] Tools are now freely available to create QMRFs.[13,14] Many aspects of the QMRF can be applied to reporting categories e.g. the endpoint, data considered, mechanistic interpretability etc., however, the basic premise of a "model" to describe is not applicable in this case. It is certainly more difficult to describe a "read-across" objectively and almost impossible to place any statistical analysis on the read-across — therefore an assessment of statistical fit and predictivity of a read-across prediction is rarely seen.

In order to provide a format to report the information from a category for read-across, following various consultations the OECD proposed the types of information that would be required to describe a category adequately.[15] This was slightly adapted in the ECHA Guidance[4] and still provides a very useful starting point for the documentation of categories. The requirements for the reporting format are summarised in Table 7.1.

7.4.1 Tools for Category Description and Prediction

The OECD QSAR Toolbox includes the full capability of producing a "QSAR Toolbox Category Report" (Chemical Category Reporting Format, CCRF) in addition to a "QSAR Toolbox Prediction Report" (QTPR). These are detailed in Step 6 of Section 4.3. The reporting capabilities are a very efficient means of collating together the process by which a category was formed in the Toolbox, the data and the prediction. The reports are editable which will be important for regulatory use. It should be noted that it is unlikely that a standard report from the Toolbox will be acceptable for regulatory use if it is simply submitted without further thought and consideration. In other words, the report will form a valuable part of the information requirement needed to complete the reporting format outlined in Table 7.1, however, it will require further expert input to provide strong and robust evidence to justify the category and the read-across prediction obtained from it. Currently, the OECD QSAR Toolbox is the only tool known to these authors that provides such a comprehensive reporting format with supporting information. However, it is likely other such tools will become available in the future as the value of this approach is more widely recognised.

7.5 Regulatory Use of Predictions

The regulatory use of categories and read-across to fill data gaps is one of the main goals of this approach. The aim here is to gain acceptance of the prediction and thus reduce, or replace, the reliance on testing. There are a number of potential uses of read-across for regulatory purposes, these include:

- prioritisation for further testing;
- classification and Labelling (C&L);
- risk assessment.

The requirements for regulatory acceptance become more stringent from prioritisation to C&L and ultimately risk assessment. In addition, the requirements will depend on the endpoint being predicted in addition to whether a prediction of a chemical being positive is "more easily acceptable" than a negative prediction.

In order for predictions of toxicity from read-across to be acceptable for regulatory purposes, a number of criteria must be fulfilled. To gain acceptance of a read-across, two basic criteria are required:

Table 7.1 Information required for reporting formats for analogue and category evaluation (adapted from OECD[13] and ECHA[4]).

Section of Reporting Format	Summary of OECD[15] and ECHA[4] Information and Guidance (adapted from the original documents).	Notes and Interpretation
1. Category definition and description of the members of the category		
1.1.a. Category hypothesis	A description of the type of structure a chemical must have to be included in the category must be given. A brief hypothesis for the formation of the category: the hypothetical relational features of the category *i.e.* the chemical similarities (analogies), purported mechanisms and trends in properties and/or activities that are thought to collectively generate an association between the members. All functional groups of the category members need to be identified. If there is a mechanistic reasoning to the category, describe the foreseen mode of action for each category member and if relevant describe the influence of the mode of administration (oral, dermal, inhalation).	This is a basic description of the category. It is vital that the information is provided to assess and confirm whether the target chemical belongs to that particular category. According to the OECD and ECHA definition, this requires very thorough description and analysis. Tools such as the OECD QSAR Toolbox will provide such information, particularly with regard to chemistry and toxicology. It is inevitable that there will be overlap between the description of a category and the definition of its applicability domain.
1.1.b. Applicability domain (AD) of the category	The applicability domain (AD) of the category is described by the inclusion and/or exclusion rules that identify the ranges of values within which reliable estimations can be made for category members. The borders of the category, and for which chemicals the category does not hold, should be indicated clearly. For example, the range of log P values or carbon chain lengths over which the category is applicable.	The definition of AD normally extends the definition of the category. This may now become quantitative in that it could include descriptors such as log P. The target chemical must fall within the stated AD of the category to ensure that it is valid.

Table 7.1 (*Continued*)

Section of Reporting Format	Summary of OECD[15] and ECHA[4] Information and Guidance (adapted from the original documents).	Notes and Interpretation
1.1 c. List of endpoints covered	The (toxicity or property) endpoints for which the category approach is applied should be listed. It should also be noted if, for some endpoints, the category approach can only be applied to a subset of the members of the category (subcategories).	This criterion appeared only in the OECD Guidance but is still relevant. Categories are sometimes (but not always) endpoint specific and this needs to be identified.
1.2. Category members	All members of the category should be described as comprehensively as possible. This should include unique identifiers including (but not limited to) CAS numbers, names and chemical structures.	The structures within the category should be noted. All structures should be recorded unambiguously identifying relevant isomerism if required. This process will implicitly assist in the description of the category and definition of its AD. It will help determine whether the target chemical is within the category. Tools such as the OECD QSAR Toolbox will perform this task for the user.
1.3. Purity/impurities	The purity/impurity profiles for each member of the category, including their likely impact on the endpoint(s) to be predicted should be provided. Any potential influence of these impurities on physico-chemical parameters, fate and (eco)toxicology, and hence on the read-across, should be identified.	Impurities are often not stated with regard to summaries of test results, therefore this will be one of the more difficult items to define and may require considerable effort and expertise. It should also be remembered that potential impurities of the target substance must be identified and their potential impact considered.

Table 7.1 (*Continued*)

Section of Reporting Format	Summary of OECD[15] and ECHA[4] Information and Guidance (adapted from the original documents).	Notes and Interpretation
2. Category justification	The category should be justified based on available experimental data (including appropriate physico-chemical data and additional test results generated for the assessment of the category). These test results should be summarised to show how they verify that the category is robust. This should include an indication of the trend(s) for each endpoint. The data should also show that functional groups not common to all the (sub)category members do not affect the anticipated toxicity. The available experimental results in the data matrix reported under 3) below should support the justification for the read-across.	This is a further description of the category but it should be more intuitive and interpretative in that it requires expert opinion to support the robustness, or otherwise, of the category. The effort required to justify a category should not be underestimated.
3. Data matrix	A matrix of data should be included that includes the members of the category associated with each endpoint. It should be constructed with the category members arranged in a suitable order (*e.g.* according to molecular weight). For example, the ordering of the members should reflect a trend or progression within the category. In each cell in the data matrix, the study result type should be indicated in the first line, *e.g.*: experimental result, experimental study planned (if applicable for *e.g.* REACH) read-across from supporting substance (structural analogue or surrogate) trend analysis (Q)SAR. If experimental results are available, the key study results should be shown in the data matrix.	The data matrix is at the heart of the reporting of the category and read-across. It is probably the first item that will be considered. It should be reported clearly, logically and unambiguously. The distinction between data should be identified *i.e.* those which are experimentally determined (and the reference for them), predictions *etc*. Tools such as the OECD QSAR Toolbox will provide this information. However, it may be the responsibility of the user to order the information correctly, annotate it succinctly *etc*.

Table 7.1 (*Continued*)

Section of Reporting Format	Summary of OECD[15] and ECHA[4] Information and Guidance (adapted from the original documents).	Notes and Interpretation
4. Conclusion(s)	Conclusions for endpoint considered for Classification and Labelling (C&L), PBT/vPvB and dose descriptor for the regulatory purposes of REACH can be provided.	This section is in the ECHA Guidance only. It provides for the possibility of making an overall conclusion with regard to the category and read-across relevant to a regulatory decision

- there is a full description of the "target" substance including its identity and purity;
- there is justification of the read-across including supporting information.

The first criterion is often overlooked, but is vital. Whilst it may seem obvious, the nature of the target substance *i.e.* the exact structure to which it relates, must be defined. The structure can be recorded through IUPAC nomenclature, SMILES or InChI strings, a drawing of the chemical structure *etc*. All areas of isomerism and the possibilities of stereoisomers or tautomers must be defined and the relative proportions stated. If the compound is formulated as a salt, or with other substances or co-solvents, then they must be defined. The reason for this careful definition is to ensure that the read-across is justifiable in terms of the applicability domain of the category.

The second criterion, the justification of the read-across, is a process of careful and expert documentation of the category and read-across. It requires evidence as defined in Section 7.4 and Table 7.1. This may be a laborious process but it is vital to demonstrate the robustness of the prediction and the reason for its acceptability.

Since 2010 much has been written about the use of read-across to provide predictions for REACH (see Table 1.3 and references contained therein as well as the recent publications from Patlewicz.).[1,2] This has involved various stakeholder initiatives providing the link between registrants (industry), academia, NGOs and ECHA. At the time of writing, such negotiations and discussions are on-going but the ultimate aim is to develop a Read-Across Assessment Framework (RAAF). One of the aims of the RAAF is to provide case studies demonstrating and illustrating good practice in category formation and read-across and, preferably, how to provide supporting evidence and document the prediction such that it may be acceptable for regulatory use. With regard to REACH, the onus is clearly on the submitter of the dossier to provide the information in a suitable format, with appropriate clarity and supported by sufficient evidence to allow regulatory agencies to accept such a prediction.

7.6 Training and Education

The premise of read-across is simple: if one molecule is similar to another then it can be assumed to have a similar activity. However, to use this principle successfully to make predictions, considerable expertise and training of non-experts is required. At the heart of *in silico* approaches is a realisation that this is a multidisciplinary methodology requiring knowledge, if not expertise, in toxicology, chemistry and possibly statistics and regulatory processes. Therefore, no one single person will enter this subject with all of the appropriate skills. The areas where training and education are required include the following (all of which are covered in this volume as well as the Guidance noted in Table 1.4);

- chemical nomenclature;
- interpretation and understanding of relevant toxicity data;
- assignment of quality to toxicity (and other) data;
- chemoinformatics tools and processes for toxicity prediction;
- tools for category formation and read-across including the process of forming a chemical category and obtaining robust and reliable toxicity data;
- reporting of read-across predictions including provision of evidence and interpretation of the relevant literature.

Currently there are excellent training materials available to support the use of the OECD QSAR Toolbox. These are available from http://www.oecd.org/env/ehs/risk-assessment/theoecdqsartoolbox.htm#Guidance_Documents_and_Training_Materials_for_Using_the_Toolbox. This includes valuable step-by-step guides and tutorials in all aspects of using the Toolbox. These will lead the user through all steps of preparing a category, performing a read-across and preparing the documentation. The training materials will not, of course, provide immediate expertise in toxicology or chemoinformatics, but they allow a scientifically-educated novice user to make a prediction and attempt to justify it.

7.7 Conclusions

Careful consideration must be given to read-across predictions of toxicity. In order for read-across to be acceptable for regulatory purposes the category and method to create it must be documented, and evidence provided to demonstrate the confidence in the read-across. There is increasing guidance available to support the definition, description and evaluation of both the categories and the read-across, although no fixed framework can be applied — it must be on a case-by-case basis and related to the endpoint. There are a number of criteria that can be considered to determine confidence, *e.g.* the strength and mechanistic relevance of the category, the numbers of analogues, quality of the toxicity data and further information from Weight of Evidence *etc*. These are all subjective and require expert opinion. When the level of confidence in a read-across has been evaluated it must be documented in a suitable reporting format. Considerable guidance and training materials are available to assist the practitioner.

Acknowledgement

Funding from the European Community's Seventh Framework Program (FP7/2007-2013) COSMOS Project under grant agreement no. 266835 and from Cosmetics Europe is gratefully acknowledged. Funding from the European Chemicals Agency (Service Contract No. ECHA/2008/20/ECA.203) is also gratefully acknowledged.

References

1. G. Patlewicz, N. Ball, E. D. Booth, E. Hulzebos, E. Zvinavashe and C. Hennes, Use of Category Approaches, Read-Across and (Q)SAR: General Considerations. *Regul. Toxicol. Pharmacol.* 2013 *In press*.
2. G. Patlewicz, D. W. Roberts, A. O. Aptula, K. Blackburn and B. Hubesch, Workshop: Use of 'Read-Across' for Chemical Safety Assessment under REACH. *Regul. Toxicol. Pharmacol.* 2013, **65**, 226.
3. European Centre for Ecotoxicology and Toxicology of Chemicals (ECETOC), *Category Approaches, Read-across, (Q)SAR. Technical Report No 116*. ECETOC, Brussels, Belgium, 2012.
4. European Chemicals Agency (ECHA), *Guidance on Information Requirements and Chemical Safety Assessment, Chapter R.6: QSARs and Grouping of Chemicals*, ECHA, Helsinki, 2008.
5. European Chemicals Agency (ECHA), *Grouping of Substances and Read-Across Approach. Part 1: Introductory Note*, ECHA, Helsinki, ECHA-13-R-02-EN, 2013.
6. European Chemicals Agency (ECHA), *Read-Across Illustrative Example. Part 2. Example 1 – Analogue Approach: Similarity Based on Breakdown Products*, ECHA, Helsinki, ECHA-13-R-03-EN, 2013.
7. M. T. D. Cronin, Biological read-across: mechanistically-based species-species and endpoint-endpoint extrapolations in *In Silico Toxicology: Principles and Applications*, ed. M. T. D. Cronin and J. C. Madden, The Royal Society of Chemistry, Cambridge, 2010, p 446.
8. M. T. D. Cronin, F. Bajot, S. J. Enoch, J. C. Madden, D. W. Roberts and J. Schwöbel, The *in chemico–in silico* interface: Challenges for integrating experimental and computational chemistry to identify toxicity. *Altern. Lab. Anim. – ATLA*, 2009, **49**, 513.
9. G. Patlewicz, O. Mekenyan, G. Dimitrova, C. Kuseva, M. Todorov, S. Stoeva, and E M. Donner, Can mutagenicity information be useful in

an Integrated Testing Strategy (ITS) for skin sensitization? *SAR QSAR Environ. Res.*, 2010, **21**, 619.
10. G. Patlewicz, M. W. Chen and C. A. Bellin, Non-Testing Approaches under REACH – Help or Hindrance? Perspectives from a Practitioner within Industry. *SAR QSAR Environ. Res.*, 2010, **22**, 67.
11. B. van Ravenzwaay, M. Herold, H. Kamp, M. D. Kapp, E. Fabian, R. Looser, G. Krennrich, W. Mellert, A. Prokoudine, V. Strauss, T. Walk and J. Wiemer, Metabolomics: A Tool for Early Detection of Toxicological Effects and an Opportunity for Biology Based Grouping of Chemicals - From QSAR to QBAR, *Mut. Res. Genet. Toxicol. Environ. Mutagen.*, 2012, **746**, 144-150
12. A. P. Worth, A. Bassan and J. de Bruijn, The Role of the European Chemicals Bureau in Promoting the Regulatory Use of (Q)SAR Methods. *SAR QSAR Environ, Res.*, 2007, **18**, 111.
13. M. Pavan and A. P. Worth, Publicly-Accessible QSAR Software Tools Developed by the Joint Research Centre. *SAR QSAR Environ, Res.*, 2008, **19**, 785.
14. A. Mostrag-Szlichtyng, J. M. Z. Comenges and A. P. Worth, Computational Toxicology at the European Commission's Joint Research Centre. *Exp. Opin. Drug Metabol. Toxicol*, 2010, **6**, 785.
15. Organisation for Economic Co-operation and Development (OECD), Environment Health and Safety Publications, Series on Testing and Assessment No. 80: *Guidance on Grouping of Chemicals*, OECD, Paris, 2007.

CHAPTER 8

The State of the Art and Future Directions of Category Formation and Read-Across for Toxicity Prediction

M. T. D. CRONIN

School of Pharmacy and Chemistry, Liverpool John Moores University, Byrom Street, Liverpool L3 3AF, England
E-mail: m.t.cronin@ljmu.ac.uk

8.1 Introduction

Read-across is the most simplistic of concepts, in that it allows an inference to be made about the properties of one object based on its similarities to another. However, the implementation of read-across for the prediction of complex phenomena, such as toxic effects, requires much greater insight. This book has brought together existing knowledge in the area of category formation and read-across. To summarise, it should now be obvious to the user that to undertake category formation to achieve read-across a number of issues must be addressed:

- the ability to "profile" a chemical for relevant structural, physico-chemical, mechanistic or metabolic information that may assist or inform a grouping.
- the ability to find chemical structures which are similar, in terms of the profiling, to the target structure.
- the ability to successfully read-across or interpolate activity.

- to interpret, find evidence to support and document the read-across to allow third parties to use, or be able to accept, a prediction based upon it.

There are an increasing number of useful techniques and approaches that allow these activities to be undertaken. However, the issues listed above become increasingly subjective, requiring more expertise, on progressing from profiling a chemical to finding evidence to substantiate a read-across prediction. The aim of this final chapter is to review the current state of the art of grouping, category formation and read-across and look to the future to describe what will be required in the short, medium and long term to increase the acceptance of read-across predictions.

8.2 Current State of the Art

Category formation to allow for read-across of toxicity is a rapidly developing area that has come a long way in the past decade. The following summarises the current status of the science through the discussion of the state of the art.

8.2.1 High Quality Tools Available to Assist the User

The area of read-across currently has a number of freely and commercially available computational tools to assist the user in grouping compounds and making read-across assessments. Such tools are described in Chapter 4 and key amongst these has been the development of the OECD QSAR Toolbox. The freely available OECD QSAR Toolbox provides access to high quality structures, profilers and databases. It allows the user to undertake read-across and trend analysis, as well as to develop simple QSAR models and provide reports to the user. The Toolbox is not the only piece of software that is available. Grouping and, to a lesser extent, Read-Across can be performed in other freely available tools *e.g.* Toxmatch and Toxtree. In addition, Chapter 4 notes a number of other freely available and commercial packages that are useful for grouping, and hence could assist in read-across. Thus, there are a number of high quality tools to assist the practitioner, at the current time there is little need to develop new tools for grouping and read-across — the focus should be on maintenance and further development (even simplification) of the existing approaches.

8.2.2 Understanding of the Processes of Grouping and Read-Across

The well documented[1-3] growth in the use of grouping and read-across is only one example of the increased confidence practitioners have in the science underpinning these approaches. Another example is the interest in Adverse Outcome Pathways (AOPs) (see Chapter 3) development and their mutual and inseparable interactions with category formation.[4] The increase in use is not only driven by necessity, *i.e.* due to the requirements of the REACH

Regulation, but also demonstrates an increasing understanding of the methods and approaches used to perform read-across. This is, no doubt, also a result of the apparent simplicity of read-across (it is worth noting that the reality is often somewhat different) and the availability of freely available tools such as the OECD QSAR Toolbox. The Toolbox in particular has received attention since its development was instigated by the OECD, with the financial backing of ECHA, and thus is perceived as already having regulatory acceptance.

8.2.3 Growth in Toxicological Databases to Support Read-Across

Readily accessible, accurate and high quality toxicity data are fundamental to read-across. However, even five years ago it was probably not possible to dream how many freely available data for read-across would now be accessible. The current data sources and means of assessing quality are described in Chapter 5, with it being clear that there is a considerable body of toxicological evidence and data available to the practitioner. These data are growing steadily as information from regulatory sources *e.g.* from REACH submission dossiers, freedom of information *etc.* is combined with automated data mining strategies. There are even the first signs that the immense data resources held by industry may be unlocked — albeit slowly and carefully.[5] At the other end of the spectrum more molecular biology data are becoming available, which will ultimately help support read-across predictions. Thus, whilst there are still insufficient data to allow for read-across predictions for all compounds for every endpoint of interest, the philosophy and motivation of data collection and retrieval is well established and supported by appropriate technology to ensure free and rapid access.

8.2.4 Status of Profilers for Category Formation

The OECD QSAR Toolbox provides a series of profilers to assist in the grouping of compounds into categories. As stated in Chapter 2 and detailed in Section 4.3 they can be defined as being mechanistic (*i.e.* potentially applicable across diverse endpoints) or endpoint specific. In addition, there are a variety of opportunities to create groupings of analogues through the use and combination of functional group-based or empirical profilers. The HESS system has identified the usefulness of developing "toxicological profilers" assisting in the creation of categories associated with specific toxicities.[6] Further development of other such toxicological profilers is anticipated. Other software such as Toxmatch allows chemicals to be grouped into categories using chemoinformatics techniques such as calculated molecular similarity. Thus, the overall picture is that for some adverse effects there is good coverage with mechanistic and/or endpoint specific profilers. Where these are not available, functional groups and/or molecular similarity may be used to create groups of analogues, but for regulatory use these may require further investigation, evidence and justification. Table 8.1 lists endpoints and provides some comment on the relative requirement for further development.

Table 8.1 Current status of the existing profilers for grouping and category formation and the requirements for their future development.

Endpoint	Current State of the Art with General Mechanistic and Endpoint Specific Profilers	Comments and Future Needs for Development
Mutagenicity and Carcinogenicity	Good coverage through a number of profilers, especially for genotoxic mechanisms	Profilers are little developed for non-genotoxic mechanisms of action
Repeated Dose Toxicity	Reasonable coverage is provided by the HESS profiler	Further development of organ level profilers is required
Reproductive/ Developmental Toxicity	Profilers are available for oestrogen receptor binding	Further development of profilers for reproductive and developmental toxicity is required
Skin Sensitisation	Good coverage through profilers relating to protein binding	Further development of profilers for skin sensitisation is not a high priority
Eye Irritation and Corrosion	Good coverage through endpoint specific profilers	Further development of profilers for eye irritation is not a high priority
Skin Irritation and Corrosion	Good coverage through endpoint specific profilers	Further development of profilers for skin irritation is not a high priority
Acute Aquatic Toxicity	Many profilers and good coverage available, especially for fish acute toxicity	Further development of profilers for acute aquatic toxicity is not a high priority, but more guidance may be required on how to use the different types of profilers
Chronic Aquatic Toxicity	No specific profilers other than those provided by ECOSAR, however, some profilers for acute aquatic toxicity may in certain circumstances be appropriate for use	Development of profilers for chronic toxicity is recommended
Endocrine Disruption	Profilers available for oestrogen receptor binding	Profilers are not currently available for *e.g.* androgen, thyroid receptors
Other toxicity endpoints	Few or no endpoint specific profilers are currently available for other toxicity endpoints. However, some general mechanistic profilers may be useful for other endpoints *e.g.* the protein binding profilers will be at least partly useful for grouping for respiratory sensitisation.	There are many areas where profilers are required

8.2.5 Prediction and Profiling of Metabolism

Metabolism and degradation are key events in toxicity and hence must be appropriately modelled when applying read-across. The prediction of metabolism is important for a number of reasons in read-across:

- Metabolites may be responsible for observed toxicity and the grouping may need to be developed around the metabolite, as opposed to the parent compound. Some profilers in the OECD QSAR Toolbox do not implicitly account for the action of important potential metabolites, whereas others do include this information. An example of this is the formation of reactive quinone compounds from poly-hydroxylated aromatic compounds.[7] This is illustrated in Table 8.2, the parent molecule (1,4-dihydroxybenzene) is not profiled as a protein binder (whilst it identified with the potential to bind to DNA by one profiler), however the predicted metabolite (benzoquinone)[8] is profiled as being a binder to DNA and protein. Such ambiguities are little understood and not easily appreciated with regard to the Toolbox and this is clearly an area for further work and clarification.
- Grouping on common metabolites is an established method of category formation.[9] Currently there are relatively few examples of how this can be undertaken, but clearly it provides a useful and mechanistically valid means to group compounds.

Table 8.2 The profiling of hydroquinone and its predicted metabolite benzoquinone illustrating the need to profile metabolites.

OECD QSAR Toolbox (ver 3.1) Profiler (see Table 4.2 for more information and definition)	Parent (Target) Compound	Predicted Metabolite from the Rat Liver S9 Metabolism Simulator (from the OECD QSAR Toolbox ver3.1)[8]
Structure	HO—⌬—OH	O=⌬=O
Name	1,4-Dihydroxybenzene (hydroquinone)	Benzoquinone
DNA Binding by OASIS	No binding	Binding by Michael Addition, Radical mechanism
DNA Binding by OECD	Binding following P450 mediated activation to a quinone	Binding by Michael Addition
Protein Binding by OASIS	No binding	Binding by Michael Addition
Protein Binding by OECD	No binding	Binding by Michael Addition

Currently, metabolic simulators are available in the OECD QSAR Toolbox, other freely available metabolic simulators include MetaPrint2D and SMARTCyp. There is also a variety of other commercial software to predict metabolism.[10] Current methods for metabolism prediction are known to have variable performance, mainly due to the fact they have been developed from data for pharmaceuticals, regardless, they should provide useful insight into the most obvious metabolites.[11] It is incorrect, however, to consider the metabolic profilers in the OECD QSAR Toolbox as a predictive platform; it is more appropriate to consider that they perform reasonably well to predict metabolites for category formation.

8.2.6 Training and Education

The rise in demand for *in silico* predictions of toxicity, *e.g.* to complete REACH dossiers, has meant that many more toxicologists and risk assessors have had to utilise category formation and read-across approaches. Many of these practitioners were previously unaccustomed to using these predictive technologies. This has inevitably meant that demand for predictions has out-stripped the available expertise. This is not to say that there is poor science being undertaken, but that issues such as evaluation of predictions for regulatory use (Chapter 7) have occasionally been poorly addressed.[1,2] On the positive side, there is much training material and other resources to help inform scientists new to this area. For example, there are a number of guidance documents from the OECD relating to the use of the OECD QSAR Toolbox and the initial case study from ECHA has recently been published.[12,13] These documents will inevitably help the user develop robust categories with which to make read-across predictions. Guidance documents are summarised in Table 1.4. In addition, it is likely that more case studies will be made available as part of the Read-Across Assessment Framework.

8.3 The Future for Category Formation and Read-Across

There is no doubt that category formation and read-across are established techniques for toxicity prediction. There are clear processes for their implementation and evidence that they will be acceptable for regulatory purposes.[1,2] There is no sign that the interest in this technique will abate. However, the category formation and read-across paradigms are not perfect and the following section identifies the areas where there is likely, and needs, to be significant development in the short, medium and long term.

8.3.1 Upkeep and Maintenance of Current Tools

The OECD QSAR Toolbox is an excellent tool for category formation and read-across and is a credit to the foresight of the OECD and developers at the

Laboratory of Mathematical Chemistry. Development of the Toolbox (Phases 2 and 3 were completed in 2012), the minimal need is to maintain the Toolbox in its current format — this is assured in the short term at least. In the long term further development may be required, not only to update profilers and databases, but also to respond to other legislative and user needs as well as exploiting future software developments. The Toolbox is only one piece of software, all other software and databases must similarly be updated and maintained to keep the impetus already created in this area.

8.3.2 Development of New Profilers

There is a clear need to continue the development of profilers for endpoints that are not yet well characterised. Table 8.1 summarises some of the main toxicity endpoints; it is clear that much effort is required in the development of profilers for chronic toxicity *e.g.* repeated dose and reproductive effects. It is true to say that many of the "easy" profilers have been developed, *e.g.* for covalent interactions with DNA and proteins, well known mechanisms of mammalian toxicity and models of non-mammalian toxicity *e.g.* fish acute toxicity. Therefore there is likely to be a much slower development of new profilers than already seen. Key areas will be in organ level toxicity and will be linked closely to the development of Adverse Outcome Pathways (see Chapter 3).

New profilers, and new ways of profiling will be required for receptor mediated interactions or molecular initiating events. Currently there is only one comprehensive profiler for a receptor mediated effect — the oestrogen receptor binding profiler in the OECD QSAR Toolbox. The new generation of profilers will not only have to address the 3-dimensional aspects of the receptor (the current profilers only include 2-dimensional properties), but also pharmacophore properties. Given the wealth of techniques in drug design for these types of endpoints,[14] there should be good possibilities to develop the next generation of profilers to address receptor mediated effects.

8.3.3 Incorporation of Toxicokinetic Information

Toxicokinetic data, particularly on metabolism and uptake, can help support a read-across prediction and support the robustness of a category.[2,9,15] One suggestion is to routinely obtain further information from any future *in vivo* studies, such as concentrations *in vivo*, possible metabolic routes *etc*.[15] The usefulness of this information to support categories derived from a common metabolite should be clear, however toxicokinetic information will also support predictions of the absence of toxicity where low bioavailability may be a useful indicator. It is clearly not possible to obtain *in vivo* toxicokinetic data for all chemicals, so greater use of *in vitro* assays and *in silico* predictions of relevant parameters is recommended.

8.3.4 Using the Adverse Outcome Pathway (AOP) Concept to Support Category Formation

As stated in the previous section, the AOP concept is becoming, or has the potential to become, central to the development of profilers for many organ level toxicities. AOPs will support the development of new categories; similarly the use of AOPs to link chemistry to a biological framework can assist in the investigation of hypotheses (for example using appropriate *in vitro* assays) relating to mechanisms of action. This cyclical process of chemistry informing intelligent testing using non-animal species relating to key events in an AOP has real potential to support the development of categories (described in detail in Section 3.4.1). There must be a realisation from experimental biologists that AOPs can support their compound selection. This will strengthen all aspects of alternatives for toxicity testing. For instance, categories supported by evidence from AOPs are, in turn, likely to be more palatable for regulatory acceptance. Therefore, linking these concepts and exploiting the knowledge of both chemists and toxicologists will be of real benefit for all parties.

8.3.5 Global Co-ordination of the Development of Adverse Outcome Pathways

In order to support the development of AOPs leading to category formation, co-ordination of these activities at a global level is required. This will ensure that effort is placed into AOPs that are relevant to on-going research activities and needs, avoiding duplication of effort. Currently AOP development is loosely co-ordinated through the OECD. This is one (of several) appropriate forums which can assist in proper review and evaluation of the AOPs.

8.3.6 Better Use of New Toxicological Data and Information Sources

There is a need to mine and use data more efficiently. This will become increasingly important in the future as data continue to be generated at an ever-increasing rate. There are already huge data resources from molecular biology and other initiatives such as ToxCast.[16] At this time we are seeing analysis of one of the most significant toxicogenomics databases, TG-GATEs, (Genomics-Assisted Toxicity Evaluation System developed by the Toxicogenomics Project in Japan).[17] The first issue will be the development of techniques to gain meaningful information from such data and relate this to mechanisms of action, hence to inform AOPs. In order to support the successful development of AOPs the hard work *e.g.* literature searching, needs to be supported and informed by these new sources of knowledge.

8.3.7 Data Quality Assessment

One of the fundamental principles of read-across is the availability of data on which to base the prediction. As noted in this chapter, there is increasing access to data from various sources. However, simply utilising these data blindly, or without consideration of their suitability, fails to address the need to assess the "quality" of the data. Methods to assess data quality are described in Chapter 5, however the current paradigm is largely based around the criteria published by Klimisch *et al.*[18] Whilst this provides a simple score for data reliability (and is very widely used) — there are acknowledged difficulties with the interpretation and application of the scheme.[19] In particular, it relies on the availability of documentation relating to Good Laboratory Practice (GLP) studies and status. Therefore, there is a need to review the requirements for data quality assessment for read-across and how this could be achieved. This requires a thorough understanding of what data quality and reliability are and how they influence predictive models. As such, new measures to provide and interpret information relating to the quality of toxicity data are needed, as discussed by Yang *et al.*[20]

8.3.8 Confidence in Predictions

In order for a prediction from read-across to be accepted, the confidence associated with the prediction must be known (see Chapter 7). Assigning confidence is a very subjective process, open to interpretation; better methods to achieve this and produce realistic and reliable estimates of confidence are required. This will not be an easy task! It will require consultation with software developers, modellers, toxicologists and users of the models in industry and regulatory agencies. At this point it may be useful for those developing methods to assign confidence to look beyond their normal areas to bring in other relevant knowledge *e.g.* from psychology, marketing, *etc.* — areas where subjective levels of confidence in an observation can be interpreted in a more objective manner.

8.3.9 Acceptance of Predictions for Regulatory Purposes

As described in Chapter 7, and noted by Patlewicz *et al.*,[1,2] one of the key stumbling blocks of the use of read-across is regulatory acceptance. It is highly desirable and certain in the (very) short term further guidance and case studies is being, and will be, provided by ECHA.[12,13] This is to be commended, as is the uptake of this guidance within the modelling community from both developers and users *i.e.* new profilers must be designed to maximise the possibility of regulatory acceptance. The logical next step is the global harmonisation of regulatory acceptance of read-across predictions and the framework to achieve this. This is achievable through the involvement of the OECD and other organisations at the global level.

8.3.10 Education and Training

There will be an on-going need for capacity building of trained experts to develop and use read-across methods, such training needs for toxicologists were reviewed by Lapenna *et al.*[21] As stated above, and in Chapter 7, there are many excellent training materials in the use of *e.g.* the OECD QSAR Toolbox, as well as courses to learn how to use the software. A further requirement is an overall strategy or framework to co-ordinate training in the broader areas of toxicology, data quality assessment, computational chemistry, regulatory acceptance *etc.* This is another area where global coordination would be beneficial.

8.4 Conclusions

Read-across as a predictive technique for chemical toxicity was little known, if at all, a decade ago. It is now a rapidly maturing technology, thanks in no small part to the accessibility of freely available software and databases. This should be a key point for future developments — uptake will be much more rapid with high quality, freely available tools. There are many areas where read-across needs to improve, such as the profilers and grouping for human health effects and the implementation of the technology for easy and effective regulatory use. The future is potentially bright for grouping, category formation and read-across. However, development of the area needs to be supported, not only by the science, but also by the uptake of the approaches by industry and the regulatory agencies. Imagination and courage are needed to take up and develop this new paradigm in hazard and risk assessment.

Acknowledgement

The funding from the European Community's 7th Framework Program (FP7/2007–2013) COSMOS Project under grant agreement no. 266835 and from Cosmetics Europe is gratefully acknowledged.

References

1. G. Patlewicz, D. W. Roberts, A. O. Aptula, K. Blackburn and B. Hubesch, Workshop: Use of 'read-across' for chemical safety assessment under REACH, *Regul. Toxicol. Pharmacol.*, 2013, **65**, 226.
2. G. Patlewicz, N. Ball, E. D. Booth, E. Hulzebos, E. Zvinavashe and C. Hennes, Use of category approaches, read-across and (Q)SAR: general considerations, *Regul. Toxicol. Pharmacol.*, 2013, *in press*.
3. H. Spielmann, U. G. Sauer and O. Mekenyan, A critical evaluation of the 2011 ECHA reports on compliance with the REACH and CLP regulations and on the use of alternatives to testing on animals for compliance with the REACH Regulation, *Altern. Lab. Anim. ATLA*, 2011, **39**, 481.

4. G. T. Ankley, R. S. Bennett, R. J. Erickson, D. J. Hoff, M. W. Hornung, R. D. Johnson, D. R. Mount, J. W. Nichols, C. L. Russom, P. K. Schmieder, J. A. Serrrano, J. E. Tietge and D. L. Villeneuve, Adverse Outcome Pathways: A conceptual framework to support ecotoxicology research and risk assessment, *Environ. Toxicol. Chem.*, 2010, **29**, 730.
5. K. Briggs, M. Cases, D. J. Heard, M. Pastor, F. Pognan, F. Sanz, C. H. Schwab, T. Steger-Hartmann, A. Sutter, D. K. Watson and J. D. Wichard, Inroads to predict *in vivo* toxicology - An introduction to the eTOX Project, *Int. J. Mol. Sci.* 2012, **13**, 3820.
6. Y. Sakuratani, H. Q. Zhang, S. Nishikawa, K. Yamazaki, T. Yamada, J. Yamada, K. Gerova, G. Chankov, O. Mekenyan and M. Hayashi, Hazard Evaluation Support System (HESS) for predicting repeated dose toxicity using toxicological categories. *SAR QSAR Environ. Res.*, **24**, 617.
7. A. O. Aptula, S. J. Enoch and D. W. Roberts, Chemical mechanisms for skin sensitization by aromatic compounds with hydroxy and amino groups, *Chem. Res. Toxicol.*, 2009, **22**, 1541.
8. O. Mekenyan, S. Dimitrov, T. Pavlov, G. Dimitrova, M. Todorov, P. Petkov and S. Kotov, Simulation of chemical metabolism for fate and hazard assessment. V. Mammalian hazard assessment, *SAR QSAR Environ. Res.*, 2012, **23**, 553.
9. European Chemicals Agency (ECHA), *Guidance on Information Requirements and Chemical Safety Assessment, Chapter R.6: QSARs and Grouping of Chemicals*, ECHA, Helsinki, 2008.
10. J. C. Madden and M. T. D. Cronin, Structure-based methods for the prediction of drug metabolism, *Expert Opin. Drug Metab. Toxicol.*, 2006, **2**, 545.
11. P. Piechota, M. T. D. Cronin, M. Hewitt and J. C. Madden, Pragmatic approaches to using computational methods to predict xenobiotic metabolism, *J. Chem. Inf. Model.*, 2013, in press.
12. European Chemicals Agency (ECHA), *Grouping of Substances and Read-Across Approach. Part 1: Introductory Note*, ECHA, Helsinki, ECHA-13-R-02-EN, 2013.
13. European Chemicals Agency (ECHA), *Read-Across Illustrative Example. Part 2. Example 1 – Analogue Approach: Similarity Based on Breakdown Products*, ECHA, Helsinki, ECHA-13-R-03-EN, 2013.
14. J. C. Madden and M. T. D. Cronin, Three-Dimensional Molecular Modelling of Receptor-Based Mechanisms in Toxicology in *In Silico Toxicology. Principles and Applications*, ed. M. T. D. Cronin and J. C. Madden, RSC Publishing, Cambridge, 2010, p. 210.
15. European Centre for Ecotoxicology and Toxicology of Chemicals (ECETOC), *Category Approaches, Read-across, (Q)SAR. Technical Report No 116*. ECETOC, Brussels, Belgium, 2012.
16. R. Kavlock, K. Chandler, K. Houck, S. Hunter, R. Judson, N, Kleinstreuer, T. Knudsen, M. Martin, S. Padilla, D. Reif, A. Richard, D. Rotroff, N. Sipesand D. Dix, Update on EPA's ToxCast Program:

Providing high throughput decision support tools for chemical risk management, *Chem. Res. Toxicol.*, 2012, **25**, 1287.
17. T. Uehara, Y. Minowa, Y. Morikawa, C. Kondo, T. Maruyama, I. Kato, N. Nakatsu, Y. Igarashi, A. Ono, H. Hayashi, K. Mitsumori, H. Yamada, Y. Ohno and T. Urushidani, Prediction model of potential hepatocarcinogenicity of rat hepatocarcinogens using a large-scale toxicogenomics database, *Toxicol. Appl. Pharmacol.*, 2011, **255**, 297.
18. H. -J. Klimisch, M. Andreae and U. Tillmann, A systematic approach for evaluating the quality of experimental toxicological and ecotoxicological data, *Regul. Toxicol. Pharmacol.*, 1997, **25**, 1.
19. K. R. Przybylak, J. C. Madden, M. T. D. Cronin and M. Hewitt, Assessing toxicological data quality: basic principles, existing schemes and current limitations, *SAR QSAR Environ. Res.*, 2012, **23**, 435.
20. L. Yang, D. Neagu, M. T. D. Cronin, M. Hewitt, S. J. Enoch, J. C. Madden and K. Przybylak, Towards a fuzzy expert system on toxicological data quality assessment, *Mol. Inf.*, 2013, **32**, 65.
21. S. Lapenna, S. Gabbert and A. Worth, Training needs for toxicity testing in the 21st century: a survey-informed analysis, *Altern. Lab Anim. ATLA*, 2012, **40**, 313.

Subject Index

β-catenin 55, 56, 57, 66

absorption studies 99
accuracy of data 115–16
acetylation of histones 55, 56, 57, 58
ACToR Aggregated Computational Toxicology Reserve (US EPA) 103, 107
acute aquatic toxicity 3, 5, 37, 135, 136
ACuteTox human toxicity data 105
acylation mechanisms 133
adequacy of data 114, 115–16
adhesion of cells 56, 66
ADME study data 99
Adverse Outcome Pathways (AOPs) 44–67
   adverse effects identification 48
   application 20
   assessment 49–50
   benefits/usefulness 51–3
   category formation 38–9
   definition 7
   development 46–9
   future directions 175
   nitrobenzene–related hemolytic anemia 145–9
   quantitative aspects 49
   recording/reporting 50–1
   SCCAs case study 53–66
   structure 46–50
AIM (Analogue Identification Methodology) software 94
aldehydes 31–2

Aldrich's Flavors and Fragrance Catalog 58
aliphatic aldehydes 31–2
alkylphenols 139–44
AMBIT database 92
Ames mutagenicity assay 78, 89, 128–32
amines 34
Analogue Identification Methodology (AIL) software 94
analogue–based categories 6, 31–2, 74, 94
anemia, hemolytic 145–9
anilines 129, 131, 132, 144, 145
animal testing 2, 11, 20, 46
   see also in vivo data
AOPs *see* Adverse Outcome Pathways (AOPs)
apical endpoints 46, 48, 66, 99
applicability domain (AD) reports 161
aquatic toxicity
   acute 3, 5, 37, 135, 136
   AOP link between MIE and *in vivo* outcome 51
   category formation case study 135–7, 138
   databases 77
   ECOSAR profiler classification 135
   endpoint specific profilers 82, 83
   short-chained carboxylic acids 57–66

Subject Index

aromatic amines 34
ARRIVE guidelines 116, 121
Ashby, J. 9
assessment
  data quality 98–124, 176
  Read-Across Assessment
    Framework 158
  regulatory acceptance 155–66
  risk 160
  similarity of chemicals 12, 13
  toxicological 2
autoxidation profilers 86

Benezra, C. 9
bioaccumulation 77, 83
bioactivity signatures 101
biodegradation 77, 80, 83
Biowin database 80, 83
bit-strings 36
bottom-up approaches 46
Bradford Hill criteria 49–50

N-cadherin 55, 56, 57
cAMP-responsive element modulator
  (CREM) 49
carboxylic acids, short-chained 53–66
carcinogenicity 77, 83, 84
case studies
  category formation 127–52
    aquatic toxicity 135–7, 138
    estrogen receptor binding 137,
      139–44
    mechanism-based 128–49
    repeated dose toxicity 144–9
    similarity-based 149–51
    skin sensitisation 132–5
    teratogenicity 150–1
  SCCAs link to developmental
    toxicity 53–66
category formation
  see also Adverse Outcome
    Pathways; computational tools
    for category formation; read-
    across
  advantages/disadvantages 15, 16

approaches 30–9
  analogue-based 6, 31–2, 74, 94
basic principles 3–5
case studies 127–52
category definition 12, 14, 76, 85,
  155–66
current understanding 169–70
future directions 173–7
history 9–11
meaning 6
process 11–15
purpose/uses 8–9, 16–17, 73–4
reporting formats 161, 162, 163
similarity definition 31
State of the Art situation 169–73
β-catenin 55, 56, 57, 66
CD34+ progenitor-derived dendritic
  cells (CD34-DC) 49
CEFIC-LRI (European Chemical
  Industry Council Long Range
  Initiative) 92
cell adhesion/motility 56, 66
cell differentiation 54, 55–7
Cell Line Activation Test 100
cell transformation database 77
central nervous system 55, 56–7
ChemBL database 104, 106, 113
Chemical Abstracts Service (CAS)
  registry numbers 108, 109, 112
chemical elements profiler 134, 137
chemical space coverage problem 34
ChemIDPlus (Advanced) database
  94, 96, 106
Chemistry Mark-up Language
  (CML) 110
chemoinformatics 6, 35–7, 74, 102,
  105
Chemoinformatics and QSAR
  Society 102, 105
ChemSpider database 93–4, 96, 103,
  106
  ID number 108, 109
1-chloro-2,4-dinitrobenzene 146, 147
class I histone deacetylase 55, 56
classification and labelling (C&L) 160

cleaned data sets 102
codes 108–13
Colipa (Cosmetics Europe) 100
completeness of data 116
computational tools for category formation 72–96
   see also OASIS
   ACToR (EPA) database 103, 107
   ACuteTox human toxicity data compilation 105
   Adverse Outcome Pathway tools/information 50
   AMBIT database 92
   Ames assay category formation case study 129
   Analog Identification Methodology (AIL) software 94
   basics 3–5
   Biowin (EPA) database 80, 83
   ChemBL database 104, 106, 113
   ChemIDPlus (Advanced) database 94, 96, 106
   ChemSpider database 93–4, 96, 103, 106, 108, 109
   COSMOS (EU) project 105
   current State of the Art 169, 170
   data quality assessment 117–21
   definition 6
   DEREK expert system 9
   DSSTox (EPA) database 107
   e-Chem (OECD) portal 103, 107
   ECOSAR (EPA) program/system 9, 135, 136
   ECOTOX (EPA) database 77, 78, 136
   EINECS (EC) 108, 109, 112
   ElementalDB app 104
   EPI Suite® database/software 78, 80, 84
   eTOX database/project 105
   HESS database/software 87–8, 144–9
   InChemicoTox 102, 106
   initial category formation 136
   ISSSTY mutagenicity database 129
   KNIME workflow software 95, 96
   Leadscope database 92–3
   MDL® toxicity database 107
   Metabolomics databases 159
   OECD QSAR Toolbox 74–87
      case studies 129, 133, 136, 140, 143, 146
   OSIRIS web tool 106
   rtER expert system 139, 140, 143
   Schwöbel database 102
   scope 2
   Terratox pesticide database 107
   TG-GATEs system 175
   TIMES mutagenicity system 80, 81, 83, 84
   Tox21 program 106
   ToxCast program 100, 103, 106
   Toxmatch program 36, 88–91, 151
   ToxRefDB 103, 107
   ToxRTool 118–21, 122
   Toxtree 91–2
   upkeep/maintenance 8, 173–4
   Vitic Nexus database 93, 103, 107
   WOMBAT/WOMBAT-PK database 107
   workflow use in read-across 95
computationally-derived chemical descriptors 113
conclusions report sections 164
confidence levels 50, 156–9, 176
consistency of data 158
cosine similarity 89
Cosmetic Regulations 11
Cosmetics Europe 100
COSMOS (EU) project 105
costs of animal testing 46
covalent bond formation
   aquatic toxicity case study 135
   DNA adducts 34, 37
   endpoints 34
   mechanism-based category formation 32, 33, 34
   Molecular Initiating Events 48
   protein binding profilers 82
   Schwöbel database 102

Cramer classification scheme 82
CREM (cAMP-responsive element modulator) 49
current State of the Art 169–73
  category formation
    profilers 170–1
    understanding 169–70
  computational tools 169
  databases growth 170
  metabolism profiling 172–3
  training/education 173
cyclohexanecarbaldehyde 136, 137, 138

data
  see also computational tools for category formation; read-across
  adequacy 114, 115–16
  Adverse Outcome Pathway development 47, 52
  category formation 8, 12, 30–1
    case studies 130, 134, 137, 140, 143–4, 146, 149
  collection 8, 12, 13–14, 104
  consistency 158
  data gaps 76, 85
  in-house data sources 101
  matrix reports 163
  on-line databases 105–7
  public sources 102–4
  quality assessment 98–124
    alternative schemes 121–2
    aspects 114–17
    chemical identifiers/codes 108–13
    computationally-derived chemical descriptors 113
    experimentally-derived data 113–17
    guidance/tools 117–21
    sources of data 101–4, 105–7
    types of data 98–101
  reporting formats 161–4
  types 37–8, 98–101

deacetylase inhibitor-induced malformations 55–7
decanoyl chloride 133, 134
9-decenoic acid 62, 66
Decision Support Systems 10
decision tree approaches 91–2
DEMETRA (EC) project 105
dendritic cell maturation 49, 77
DEREK expert system 9
descriptors, chemical 108–13
developmental toxicity
  analogue-based categories 32
  category formation case studies 150–1
  deacetylase inhibitors 55–7
  developmental hazard index 58, 59–60, 62, 63–5
  mechanistic overview 53–5
  multidimensional pathways 45
  short-chained carboxylic acids 53–66
  Toxmatch software use 36
DHI (developmental hazard index) 58, 59–60, 62, 63–5
differentiation of cells 54, 55–7
3,4-dimethyl-1-pentanol 3, 5
dimethylanilines 131, 132
4-(2,2-dimethylpentyl)phenol 139, 141
Direct Peptide Reactivity Assay (DPRA) 81, 100
DNA adducts
  see also mutagenicity
  DNA binding by OASIS profiler 129
  DNA binding by OECD profiler 129
  in silico-based profilers development 37
  OECD QSAR Toolbox example 76
  structural alerts 34
documentation
  see also literature
  AOP reporting 50–1
  OECD QSAR Toolbox 86–7

regulatory acceptance 156, 159–60, 161–4
reporting formats 50, 159, 160, 161–4
toxicity prediction 12, 14, 15
DPRA (Direct Peptide Reactivity Assay) 81, 100
DSSTox (EPA) database 107
Dupuis, G. 9

e-Chemportal (OECD) 103, 107
EC-JRC (European Commission Joint Research Centre) 19, 88–91, 92
ECETOC (European Centre for Ecotoxicology and Toxicology of Chemicals) 18, 133, 136, 158, 159
ECHA see European Chemicals Agency (ECHA)
ECOSAR (EPA) program/system 9, 135, 136
ECOTOX (EPA) database 77, 78
ECVAM (European Centre for the Validation of Alternative Methods) 118
education for read-across 165, 173, 177
effects identification 12, 13
EINECS (European INventory of Existing Commercial chemical substances) number 108, 109, 112
electrophilic mechanisms 34, 129
ElementalDB similarity-based search app 104
embryological toxicity see developmental toxicity
empiric profilers 84–5, 130, 133–4, 136–7, 143
endpoints
 Adverse Outcome Pathways 46, 48
 category formation 73
 chemical similarity definition 31
 identification for prediction 12, 13
 list for reporting format 162
 OECD QSAR Toolbox 76, 79, 82–4
 profilers 82–4, 128–49
 regulatory guidance 17
EPISUITE (EPI Suite®) database 78, 80, 84
estrogen receptor binding 35
 category formation case study 137, 139–44
 databases 77, 78
 endpoint specific profiler 84
 new profilers 174
 profiler case study 139, 140, 141–2, 143
2-ethylhexanal 138
2-(3-ethylphenyl)oxirane 3, 4
ethynodiol 151
ethynodiol diacetate 150, 151
eTOX database/project 105
Euclidean distance 89
European Centre for Ecotoxicology and Toxicology of Chemicals (ECETOC) 18, 133, 136, 158, 159
European Centre for the Validation of Alternative Methods (ECVAM) 118
European Chemical Industry Council Long Range Initiative (CEFIC-LRI) 92
European Chemicals Agency (ECHA)
 definitions 6
 non-confidential regulatory submissions 102
 OECD QSAR Toolbox development 74
 QSAR regulatory acceptibility 10–11
 relevance of data guidelines 115
 report 16
 reporting guidance 160
European Commission

Subject Index

category formation/read-across 9–11
ChEMBL database 104, 106, 113
COSMOS project 105
DEMETRA pesticides project 105
Joint Research Centre (EC JRC) 19, 88–92
European INventory of Existing Commercial chemical substances (EINECS) number 108, 109, 112
expert systems 7, 9, 122, 139, 140, 143
exposure assessment 2
exposure-based waiving 10
eye irritation 83

fathead minnow (*Pimephales promelas*) 3, 5, 135, 137, 138
FETAX (Frog Embryo Teratogenesis Assay - *Xenopus*) 59–60
fingerprint methods 36–7
fish acute toxicity 3, 5, 37, 135, 136
fitness-for-purpose 108–24
formats for reporting 50, 159, 160, 161–4
Frog Embryo Teratogenesis Assay - *Xenopus* (FETAX) 59–60
frog (*Xenopus laevis*) 57, 58, 59–60
functional groups 31–2, 39, 84
  see also organic functional group profiler
future directions 173–7
  Adverse Outcome Pathway use 175
  computational tools upkeep/maintenance 173–4
  confidence levels 176
  data quality assessment 176
  global co-ordination 175
  new profiler development 174
  regulatory acceptance 176
  toxicokinetic data incorporation 174
fuzzy expert systems 122

gastrulation 54, 55
gene mutations *see* mutagenicity

Genomics-Assisted Toxicity Evaluation System (TG-GATEs) 175
genotoxicity databases 78
global aspects 103, 107, 175, 176
Good Laboratory Practice (GLP) 121–2, 176
good practice 121–2, 164–5
grouping *see* category formation

Hashed InChi (InChi Keys) 111, 112
Hazard Evaluation Support System (HESS) software 87–8, 144–9
hCLAT (human Cell Line Activation Test) 100
Hellinger distance 89
hemolytic anemia 145–9
heptanal 138
4-tert-heptylphenol 142
4-heptylphenol 142
HESS (Hazard Evaluation Support System) software 87–8, 144–9
hexadecanoyl chloride 134
hexanal 138
4-tert-hexylphenol 141
4-hexylphenol 142
high throughput screening (HTS) 52, 53, 73, 100
histogenesis 54
histone acetylation 55, 56, 57, 58
history 2, 9–11
Hodgkin–Richards index 89
human Cell Line Activation Test (hCLAT) 100
hydrolysis profilers 81, 86
hydrophobicity 146
HYDROWIN software 81

identifiers, chemical 108–13, 164
impurities 158, 162
in chemico assays 37–8, 39, 47
in silico see computational...
in vitro data
  Adverse Outcome Pathway development 47

profiler development for category formation 37, 39
reliability 114
ToxRTool questions posed 118, 120
types 99
in vivo data
    Adverse Outcome Pathway development 47, 51, 52
    Molecular Initiating Event link to *in vivo* endpoints 44
    profiler development for category formation 37, 39
    reliability 114
    ToxRTool questions posed 118, 119
in-chemico assays 34
InChemicoTox project 102, 106
InChI (IUPAC International Chemical Identifier) codes 108, 110, 112
InChiKeys (Hashed InChi) 111, 112
in-house data sources 101
initial category formation case studies 136, 139, 140, 143, 145–6
inputs to OECD QSAR Toolbox 75, 76
Integrated Testing Strategies (ITS) 10
intermediate AOP events 49
international collaborative projects 103
International Union of Pure and Applied Chemistry (IUPAC) names 108, 109, 112
ionisation 12, 81
ISSSTY bacterial mutagenicity database 129
ITS (Integrated Testing Strategies) 10
IUPAC (International Union of Pure and Applied Chemistry)
    names 108, 109, 112
    InChI codes 108, 110, 112

journal articles 102, 105
    see also literature

JRC (European Commission Joint Research Centre) 19, 88–91, 92

keratinocyte gene expression 83
Klimisch scoring scheme 117–18, 121, 176
KNIME workflow software 95, 96

Leadscope database 92–3
legislation
    Good Laboratory Practice 121–2, 164–5, 176
    REACH regulation 30, 115, 164, 169–70
    RAAF 158, 164–5
    safety assessment requirements 8
    standards 47–8, 116
Lien, E.J. 9
line notations 109–10
literature
    Adverse Outcome Pathway key events recognition 49
    journal articles 102, 105
    living documents 50
    regulatory guidance for read-across 16–19
        ECHA report 16
        key documents 18–19
    reviews/summaries for key data 105
liver toxicity 39, 145–9
living documents 50
log P (partition coefficient) 3, 4, 8, 141–3
Lowest Observed Effect Level (LOEL) 101, 146–9
lynestrenol 151

MDL® toxicity database 107
mechanism of action (MOA)
    case-studies 128–49
    category formation 32–5, 39, 73
    OECD QSAR Toolbox category definition step 85

Subject Index

profilers 80–2
metabolic pathways
  databases 102–3
  Metabolomics databases 159
  OECD QSAR Toolbox 76, 86
  profilers 172–3
3-methyl-4-pentenoic acid 62, 64
methylanilines 129, 130, 131
2-methylbutanal 138
2-methylpentanal 138
methylpentanoic acids 62, 63, 64
micronucleus mutagenicity databases 78
MIEs see Molecular Initiating Events (MIEs)
mitochondrial membranes 48
MOA see mechanism of action (MOA)
molecular descriptors 7
Molecular Initiating Events (MIEs)
  Adverse Outcome Pathways 44–8, 51, 52
    category formation 38
    definition 48
    key events recognition 49
    link 51
  chemical similarity definition 31
  covalent bond formation endpoints 34
  definition 7
  mechanism-based category formation 32, 33, 34
  mechanism-based primary profilers 128, 129, 130
  non-covalent endpoints 35
  similarity-based category formation case studies 149
  skin sensitisation category formation case study 135
molecular modelling 7
monocyte chemotactic protein-1 receptor (CCR2) 49
mouse embryos 57
multi-dimensional pathways 45, 100

MUSST (Myeloid U937 Skin Sensitisation Test) 100
mutagenicity
  Ames assay 78, 89, 128–32
  databases 78
  endpoint specific profilers 83
  2-(3-ethylphenyl)oxirane 3, 4
  ISSSTY database 129
  read-across predictions 3, 4
  Salmonella typhimurium 3, 4
  supermolecule (Ashby/Tennant) 9
  TIMES system 80, 81, 83, 84
  Weight-of-Evidence approaches 159
Myeloid U937 Skin Sensitisation Test (MUSST) 100

N-cadherin 55, 56, 57
National Institute of Technology and Evaluation (NITE) 144
nephrotoxicity 51
neural crest 56–7
neural tube defects 55, 56, 57
neurotoxicity 51, 55, 56–7
NITE (National Institute of Technology and Evaluation) 144
nitrenium ions 34
nitrobenzenes 144–9
No Observed (Adverse) Effect Level (NO(A)EL) 101
nomenclature, chemical 108, 109
non-confidential regulatory submissions 102
non-covalent bonds 33, 34, 48
nonanoyl chloride 134
norethindrone 151
number of compounds in group 157, 158

OASIS
  acute aquatic toxicity mechanism of action profiler 135, 136
  aquatic toxicity database 136

estrogen receptor binding affinity database 140, 143
genotoxicity database 129
protein binding by OASIS mechanistic profiler 132, 133, 135, 136
octadecanoyl chloride 134
4-octylphenol 142
OECD *see* Organisation for Economic Development (OECD)
oestrogen *see* estrogen
on-line databases 105–7
organic functional group profilers 134, 137, 139, 140, 143
  see also functional groups
Organisation for Economic Development (OECD)
  definitions 6, 30
  e-Chemportal 103, 107
  GD 34 validation guideline 116
  key documents 18
  OECD QSAR Toolbox 3, 4, 5, 74–87, 106
    category formation 31, 127–52
    current State of the Art 169, 170, 172, 173–4
    databases available 77–9
    robustness of categories 157
    training/education for use 165
    versions release 10
    workflow 11, 12, 14–15, 75–87
  Principles for the Validation of (Q)SARs 159
  Screening Information Data sets dossiers 105
  Test Guideline Programme 53
  validation principles 10
organogenesis 54
OSIRIS 106
outcomes *see* Adverse Outcome Pathways (AOPs); endpoints

paracetamol 112
partition coefficient (log P) 3, 4, 8, 141–3

pathway-based approaches 45
  see also Adverse Outcome Pathways (AOPs); Molecular Initiating Events (MIEs)
peptide reactivity assay 81, 100
4-phenethylphenol 141
phenolic weak acid uncouplers 48
pictorial representation 108, 109, 112
Pimephales promelas 3, 5
  assay 135, 137, 138
planning processes 156
potency index (PI) 58, 61, 62, 63–5
potential toxicity 73
primary profilers 128–49
prioritisation for further testing 160
profilers
  Adverse Outcome Pathway category formation 38
  definition 6
  mechanism-based category formation case studies 128–49
  mechanism-based structural alerts 33
  new developments 174
  OECD QSAR Toolbox 76, 79, 80–5, 86
2-propylpentanoic acid (valproic acid) 55–7
protein binding 82, 132, 133, 135, 136
public data sources 102–4
publications *see* literature

QSARs *see* Quantitative Structure–Activity Relationships (QSARs)
quality assessment 98–124, 176
quantitative aspects of Adverse Outcome Pathways 49
Quantitative Structure–Activity Relationships (QSARs) 5–7, 10
  see also Organisation for Economic Development (OECD), OECD QSAR Toolbox

QSAR Model Reporting Format 159, 160

RAAF (Read-Across Assessment Framework) 158, 164–5
rainbow trout estrogen receptor binding (rtER) expert system 139, 140, 143
rat model 58, 61, 86, 88–9
REACH regulation 30, 115, 164, 169–70
   Read-Across Assessment Framework 158, 164–5
read-across
   see also workflows for read-across
   accepting/rejecting 14
   Assessment Framework (RAAF) 158, 164–5
   basics 3–5, 6–7
   current understanding 169–70
   definition 6
   future directions 173–7
   history 9–11
   process 11–15
   purpose 8–9
   regulatory acceptance 16–19, 155–66
   short-chained carboxylic acids 62, 63–5
   State of the Art situation 169–73
   tools for category formation 72–96
   training/education 165
   uses 16–17
Read-Across Assessment Framework (RAAF) 158, 164–5
receptor-mediated effects 137, 139–44, 174
regulatory acceptance 155–66
   achievement 8
   Adverse Outcome Pathway applications 53
   computational tools 160
   confidence 156–9
   definition 6
   future 176

predictions 157–60, 161–5
Qualitative Structure–Activity Relationships 10–11
training/education 165
Weight-of-Evidence 159
regulatory non-confidential submissions 102
relevance of data 114, 115
reliability of data 114, 115
repeated dose toxicity 78, 79, 87–8, 144, 145, 146
reporting *see* documentation
reproducibility of data 116
reproductive toxicity 51
respiratory sensitisation 34, 37, 51
risk
   assessment 160
   definition for safety assurance of chemicals 1–2
   developmental hazard index 58, 59–60, 62, 63–5
   HESS system 87–8, 144–9
robustness of categories 156–7
rtER (rainbow trout estrogen receptor) expert system 139, 140, 143

safe chemicals assurance 1–7, 8
Salmonella typhimurium 3, 4, 129, 130, 131, 132
Schiff base formation 136
Schwöbel database 102
Screening Information Data sets (SIDS) dossiers 105
secondary profilers 128–49
Setubal Workshop 10
short-chained carboxylic acids (SCCAs) 53–66
signatures of toxicity 101
similarity of chemicals
   case studies 149–51
   category robustness 156–7
   chemoinformatics 35–7, 39
   definition 6, 31
   identification/assessment 12, 13

Tanimoto coefficient 36, 89, 92, 151
Toxmatch program 88–91
Simplified Molecular Input Line Entry System (SMILES) strings 75, 108, 110, 112
size of categories 157, 158
skin metabolism 86
skin sensitisation
　Adverse Outcome Pathway 49, 51, 52
　category formation case study 132–5
　covalent bond formation endpoints 34
　data quality assessment 100
　databases 78
　endpoint specific profilers 84
　multidimensional pathways 45
　OECD QSAR Toolbox example 76
　profiler development for category formation 37, 39
　Toxmatch software use 36
　Weight-of-Evidence approaches 159
SMILES (Simplified Molecular Input Line Entry System) strings 75, 108, 110, 112
software *see* computational tools for category formation
Solna Principles 116–17
standards 47–8, 116, 121–2, 176
State of the Art situation 169–73
structural alerts
　AOP development 47
　covalent bond formation endpoints as Molecular Initiating Events 34
　definition 6
　mechanism-based category formation 32–5
　non-covalent bonding endpoints category formation 35
　profiler types 37, 128
　short-chain carboxylic acids 61

toxicity/nontoxicity indication 33
Toxtree toxicity endpoints 91
two-dimensional 35
structure representation 108–13
Structure–Activity Relationships (SARs) 6
　see also Quantitative Structure–Activity Relationships (QSARs)
sub-profiling 38

TA1537 *Salmonella* strain 130, 131, 132
Tanimoto coefficient (of similarity) 36, 89, 92, 151
target chemical identification 11–13, 164
templates for reports 50, 161–4
Tennant, R.W. 9
teratogenicity *see* developmental toxicity
terminology 47–8, 108–13, 164
TerraTox database 107
TG-GATEs (Genomics-Assisted Toxicity Evaluation System) 175
three-dimensional aspects of new profilers 174
TIMES system 80, 81, 83, 84
tools *see* computational tools for category formation
top-down approaches 46
Tox21 program 106
ToxCast program 100, 103, 106
toxicity data *see* data
toxicity endpoints *see* endpoints
toxicity signatures 101
toxicity types/categories 87
Toxmatch program 36, 88–91, 151
ToxRefDB (Toxicity Reference Database) 103, 107
ToxRTool (Toxicological data Reliability Assessment Tool) 118–21, 122
training set data 88
training/education 165, 173, 177

Subject Index

transformation of cells 77
two-dimensional structural alerts 35, 139, 150–1

uncertainty of data 115–16
unique identifiers 108, 109
United States
   Environmental Protection Agency
      ACToR database 103, 107
      BLOWIN® exposure assessment tool 80, 83
      DSSTox database 107
      ECOSAR program/system 9, 135, 136
      ECOTOX aquatic toxicity database 77, 78, 136
      rtER expert system 139, 140, 143
   teratogenicity 150, 151

validation of data 10, 116–17, 155–66
valproic acid 55–7
variation sources in data 114
Verhaar profiler 136
vertebrate embryo neural crest 56–7
Vitic Nexus database 93, 103, 107
VITROSENS test 49

weak acids 48, 55–7
Weight-of-Evidence (WoE) approaches 49–50, 66, 123, 124, 159
Wnt/β-catenin signalling pathway 55, 56, 57
WOMBAT/WOMBAT-PK database 107
workflows for read-across
   Ames assay 129–32
   aquatic toxicity 137
   estrogen receptor binding 139–40, 143–4
   KNIME workflow software 95, 96
   OECD QSAR Toolbox 75–87
   repeated dose toxicity 145–6, 149
   skin sensitisation 133–5
   use in read-across 95

Xenopus laevis 57, 58, 59–60